GOLD
IN THE ANCIENT WORLD

HOW IT INFLUENCED CIVILIZATION

JENIFER G. MARX

All photos courtesy of Jenifer and Robert F. Marx.

JANICE HARVIS
1939–2008

This book is dedicated to a person who was a real inspiration to many, many people. She encouraged and, at the same time, lifted up the spirits of those whom she came in contact. For centuries, gold has been the criteria for excellence and no better person could be associated with this precious mineral. Her standards, carriage and bearing all represented the highest quality of life. She often said that it took seven generations to make a lady, and Janice was a lady.

ISBN 0-9818991-1-0

Gold in the Ancient World
How it Influenced Civilization

Dewey Reference Number: 972.9

Library of Congress Cataloging-in-Publication Data

Marx, Jenifer.
Gold in the ancient world : how it influenced civilization / Jenifer G. Marx.
p. cm.
ISBN 978-0-9818991-4-5 (alk. paper)
1. Gold--History. 2. Gold in art--History. 3. Civilization, Ancient. I. Title.
GT5170.M37 2009
669'.22093--dc22
2009000247

Published by RAM Books
Garland, Texas
A Division of Garrett Metal Detectors

Contents

INTRODUCTION

Gold, the precious metal that has fascinated mankind since the dawn of time, has played a role in my life and in the development of Garrett Metal Detectors. I have spent countless hours in the field recovering gold nuggets and flakes that now glitter in my safety deposit box. Some of the earliest books published by our RAM subsidiary explained how Roy Lagal and I used pans and detectors to find gold, even as we coined the term "electronic prospecting."

And, I know for a fact that many of our customers during the past 45 years have purchased Garrett Detectors *specifically to search for gold.* This has been especially true during the times in year past when the price of gold soared as high as $800 an ounce. It is a fact that orders for our detectors literally overwhelmed the factory, and we were weeks behind in delivering product. I also believe that although most detector buyers don't plan to use their metal detector to find gold, the thought remains in the back of their minds that these miraculous instruments might *someday* discover gold for them.

Amateur treasure hunters and professional prospectors alike know that gold is a remarkable and precious element closely associated with man's yearning for riches throughout the ages. In fact, gold is mentioned more than 400 times in the Bible. Men and women of all walks of life have always loved gold and desired it both as a decorative object and for its monetary value..

Today, there is a renewed excitement to prospecting for gold since the metal's value recently surpassed $1,000 per ounce. I find it fascinating that this element is still earnestly sought—and will continue to be—after so many thousands of years.

The history of gold tells a fascinating story of the many ways in which it has helped to shape our modern society...so fascinating that RAM went beyond its usual "how-to" style in 1996 to publish *Gold of the Americas*. In this delightful book, popular author Jenifer Marx explains in interesting detail how the lust for gold was essentially responsible for both the discovery of America itself and then the western settlement of the United States.

It's been my great pleasure to know Jenifer and her famous marine archaeologist husband, Sir Robert Marx, for many years. The amount of research they put into their treasure expeditions is extraordinary. Bob has used Garrett's *Sea Hunter* metal detectors literally all over the world to assist him in discovering countless riches beneath the seas. He first met Jenifer on one of his archaeological ventures, and she has since joined him on some of his journeys to bring golden treasures from the ocean's depths.

Her life's work in this field and training in research have given Jenifer great expertise concerning gold's history. Our 1996 volume offered proof. Now Jenifer has dug back far beyond her history of gold in the Americas. *Gold in the Ancient World* paints a broad picture of adventure, artistry and greed as it explores achievements resulting from man's love of gold...wars and rituals that have surrounded the precious metal for more than fifty centuries.

As this book is published, golden treasures from the tomb of King Tut are once again touring the United States. Remarkable treasures such as these sometimes seem to be the only legacy of long-vanished civilizations. Fortunately, *Gold in the Ancient World* offers an opportunity to learn from factual history how gold built up and tore down vast empires. Illustrations in the book depict many of the golden treasures that drove mankind to marvelous feats over the years.

I truly believe that anyone interested in gold, whether an active prospector or not, will benefit from reading this history of the metal that is being sought so enthusiastically today with 21st-century detectors. Just as a coin hunter's library is incomplete without coin guides, so too is the prospector's bookshelf lacking without this detailed work on the how and why of gold's influence on mankind.

I know that you'll enjoy this book as I continually do because of its fascinating history and photo documentation. I hope to see you in the gold fields...

Charles Garrett

CHAPTER 1

GIFTS FROM THE SUN

Gold is remarkable stuff. Since the calm of history it has played a prominent part in human affairs. The dazzling yellow metal has fueled the highest artistic achievements for the past six thousand years providing a superb medium for craftsmen whose work reflects the changing character of passing time. In the ancient world men and women honored their gods, their dead and themselves with beautiful gifts of lustrous gold. The golden artifacts exhumed from the darkness of the past by archaeologists or tomb robbers are sometimes the only legacy of a long vanished civilization. They glow at us from museum cases amazing in their loveliness. They evoke feelings in us of wonder and kinship with people of the distant past.

The magnificent artworks of Ur, of Egypt or the proud horsemen of the Russian steppes are enduring and tantalizing witnesses to a time when men lived intimately with metal sacred to the sun, the oldest of the gods. From earliest times gold was symbolic of solar light, which was recognized by the ancients as the life's giving principle. Gold thus was the most fitting offering to the gods and to rulers who were identified with the supernatural. It represented divinity, authority and security and gave great power to those who had it. At the same time it gave pleasure to the skilled hands that shaped it and brought suffering to those who mined or gathered it.

"What dost thou not compel the human heart to do, accursed greed for gold," the Roman poet Virgil wrote in the first century

B.C. when already countless thousands of slaves had died hacking it out of Egyptian mines for the pharaohs, or washing it from Spanish rivers for foreign overlords. Possession of the rare and precious metal throughout much of history has been the exclusive right of the ruling classes. A substance that is scarce, unchanging, lovely, and easily shaped, the yellow metal has had a mystical but very strong association with power—both divine and temporal.

Religion, politics, economics, romance and magic have all been subject to the influence of mankind's oldest and least perishable treasure. It has been the stuff of myth, creation and destruction until modern times. No other substance has captured a man's imaginations and fired his passions as gold has. The metal which has inspired magnificent dreams and provoked bloody wars retains, even in our time, its ages-old appeal. Yet, the role of gold in the future is uncertain. Men no longer pay golden homage to gods and kings. The dead are not furnished with gleaming gold in the darkness of their tombs. The royal metal has already been dethroned as king in global economics. And, in the shrunken world of satellite communications and computerized technology where every mountain peak has been scaled and every jungle penetrated, there is little place for the dreams of the gold-seeker.

But even now, despite the complexity of large-scale production and marketing of gold from South Africa, the Soviet Union, and other areas, lone prospectors still hunt for a bonanza. The men who might have trekked to the American West after 1849, to the Klondike or Australia later in the 19th century, now follow the meanders of streams in Honduras or hack their way through tangled tropical growth in South American rainforests. No cure has ever been found for gold fever.

It seems hard to imagine that if all the gold that has ever been produced were amassed in one place it would fit into one supertanker. The demand for gold has always been in excess of availability since the alluvial deposits along the Nile River were worked some seven thousand years ago. The hunger for gold paradoxically seems to have increased even as it was fed. In the first century B.C.

the king of gold-rich Parthia practiced a literal and horrible justice to punish Marcus Crassus, a captured Roman general who had invaded Parthia for its gold. The king fed the Roman's hunger with molten gold poured down his throat. Counterfeiters in medieval Russia were similarly dealt with when sentenced for falsifying gold coins—a practice born with coinage in early 6th century B.C.

Few people today are aware of living with gold. Mass production has diminished the once sacred function of the goldsmith and lowered the quality of gold ornament and jewelry. Gold coins are collected but seldom circulated. Great treasures lie in dusty displays or museum storerooms, and are only occasionally put in the spotlight. Huge piles of gold bullion, still the bulwark of national economies and the payment of last resort, lie in dank underground vaults where their sheen is unappreciated. Today, gold is used in thousands of invisible ways. The market for gold continues to expand as new applications are found in fields ranging from medicine to the space industry.

Economists strive to reduce gold to the status of a trading commodity on a par with copper or coffee. But more than six thousand years of experience have proved that gold's aura can't be so easily dispelled. Gold is forever, defying political and economic upheavals as no paper money can. It seems unlikely that mankind will cease to have a very special feeling for the precious and beautiful substance that has played such a complex role in the human drama. Whatever its part – curse or cornerstone, gold has always emerged gleaming and in no way diminished.

The most highly prized of metals has almost always been produced with the cheapest of labor. Legions of miners have been sacrificed to gold from the scorched deserts of Africa to the frigid wastes of Stalin's 20th century Siberian mines. Even now in South Africa, which produces about two-thirds of the world's gold, mines claim their victims. Gold lust is so compelling that unknown numbers have sacrificed themselves, plunging into immense jungles or the ocean depths in search of the elusive dream. The compulsion that drove Jason and his Argonauts after the Golden Fleece and propelled restless gold seekers back and forth across the globe in the 19th century still sends determined dreamers in search of lost mines and the legendary golden cities of the ancients.

Gold—buried, sunken and entombed—lures treasure hunters to every corner of the world. In the towering Andes they hunt for the missing treasure of the Incas which evaded Spanish capture. In central Italy grave robbers probe the ground for golden relics of the mysterious Etruscans. Gold hidden away during the upheavals which convulsed much of Asia Minor in ancient times is a treasure many seek. The splendid Lydian gold hoard unearthed in Turkey, smuggled to New York, and bought by the Metropolitan Museum of Art was one such shining prize. The waters of the world also beckon treasure hunters who scour lakes, rivers and seas for sunken ships laden with gold, for sacrificial golden idols and offerings. The most zealously sought treasure of them all is probably the sunken continent of Atlantis whose golden spires allegedly gleam beneath the sea.

Because gold endures, it accumulates. The greater part of the gold that passed through the hands of the Mesopotamians, Egyptians, Persians, Scythians, Incas and Aztecs, to name some who held the yellow metal sacred, is still present. Some estimates run as high as 80%. Of course, very little of that is in its original form. Gold is almost indestructible but it has an Achilles' heel. It is very soft—a property both a curse and a blessing. Because it is malleable, goldsmiths of great talent have regarded it as the ideal medium in which to express the focus and fashion of their particular time.

But the fact that gold is so easily worked and so easily melted meant that when fashion changed, women could bring outmoded ornaments or accessories to goldsmiths who refashioned them according to current taste. The gold artifacts of the ancient world which have survived came chiefly from tombs and buried hoards hidden in the face of invasion, battle or other disaster. In times of war or economic hardship or even cultural revolutions such as Henry VIII's rupture with the Pope, gold had a way of disappearing either into the ground or into the melting pot.

The king of metals has passed down through the ages in and out of fire and the sensitive hands of goldsmiths and coiners, undergoing one transformation after another. A gold nugget washed from the sands of the Nile, made into a pharaoh's jewel and entombed with him could today be part of an astronaut's visor. It might conceivably have been robbed from the tomb and traveled through the centuries as a small ingot, a Roman matron's hair ornament, a coin and in other uses until it became part of a 20th century space odyssey hurtling toward the heavens from which the ancients believed it came.

Man has yet to find a universally acceptable substitute for gold as the ultimate store of value. Even in our uncertain times the durability of gold represents a constant, defying both inflation and depression. Hailed by the Greek playwright Euripides as "the most beautiful thing a man can take in his hand," gold still symbolizes wealth, prestige, security and love. What is this inert mineral which has held men in its thrall esince the first prehistoric hunter picked up a nugget, tumbled smooth in a mountain stream, and marveled at its luster as it lay in his palm? Gold is beautiful. No one disagrees. However, aside from its natural loveliness, it has no intrinsic value. To a man lost without food or water in the scorching sands of the gold-bearing Arabian Desert all the gold of Solomon would be of no use. Nor would it help him if he were shipwrecked on a deserted island or found himself in any of the cultures where gold is not prized. In the Solomon Islands of the Pacific, for example, a young man who wants to purchase a wife will have better luck with 500 porpoise teeth than a handful of gold. In Samoa mats, not gold, are accepted as currency. On the island of Yep a man's wealth is reckoned in terms of huge pierced stone discs. But where gold has been available, the metal prevails as the most trusted store of wealth. No other substance can match its long history of service in economics or boast such a wide range of associations and psychological appeal.

"Gold is the child of Zeus" wrote Pindar in the 5th century B.C., "neither moth nor rust devour it; but the mind of man is devoured by this supreme possession." The statements the Greek poet made about gold are keys to its exalted status. Zeus, the supreme deity of the Greeks, was originally a sun god. Mankind's earliest gods were identified with the sun, and from the beginning of human history gold was also identified with the sun as heavenly metal. It was regarded as having come from the fiery orb, the source of light, heat and life. As the offspring of the sun, gold was worshipped as the highest form of matter, and cosmic powers were attributed to it.

Gold was the material symbol of all that was eternally pure, shining and perfect to the Chinese, the Hindus, and the Egyptians. Ancient Mongol scrolls describe the continents as the Golden Surface. The Chinese believed the world's center was on Mount Sumeru on a circle of gold around which the sun and moon revolved. Their creation myth is one of many in which gold appears. From Mount Sumeru four great rivers flowed including the Indus, which flowed southwest from the mouth of a golden elephant.

While goldsmiths crafted physical gold into objects of lasting beauty, philosophers of East and West worked with metaphysical gold, which not only represented the highest spiritual qualities but was indispensable to bards and writers as an ingredient in poems, proverbs and romances. The golden metaphor is a two edged sword. As well as being the epithet for all things lovely and true, gold represents the erosion of humanity and principle – the lust, greed and corruption which poison men's souls.

"In consequence of gold there are no brothers, no parents; but wars and murders do arise from it and for it we lovers are bought and sold," lamented an ancient Greek. Two thousand years later Shakespeare wrote, "How quickly nature falls into revolt when gold becomes her object!" It is, of course, the minds of men that gave gold the dual powers to inspire and corrupt. Gold has no inherent powers of creation or destruction. It isn't even essential to survival, yet it has created empires and destroyed whole civilizations.

What is gold? It is a chemical element—a heavy, dense, soft yellow metal. Each of the known elements has been assigned a number according to the complexity of its atoms beginning with hydrogen, the simplest, at number 1. Gold at 79 ranks between platinum and mercury. Gold has unrivaled physical and chemical properties, which account as much for its ancient association with the mystical as with valuable new applications in our electronic space age.

The most obvious appeal of gold is its shining natural beauty. But its most significant property is what medieval alchemists, who labored to turn inferior metals into gold, called nobility. As Pindar noted, the heavenly metal is impervious to the ravages of time and the elements. Gold is chemically inactive; neither air, moisture nor common acids and alkalis can mar its loveliness. A golden surface doesn't dim, mildew, rust or corrode, which is why the splendid treasures of the boy king Tutankhamen concealed in his tomb for more than 3,000 years are as dazzling as the day they were made, long after most other artifacts would have decayed.

The first Egyptian hieroglyph for gold was a stylized necklace such as the ornate gold collars which have been found in royal tombs. Later, gold was represented by a circle in Egypt and elsewhere in recognition of the relationship between the sun and the divine metal. The sun symbol lasted into the 18th century in the Western world when it was replaced with Au, derived from the Latin for radiant dawn. Aurum, the Latin word, traces its origin to a Sanskrit meaning "to burn." The English word "gold" is related to the oldest words for yellow in the Indo-European languages. It is also linked to the ancient words for Yule. At annual ceremonies the Yule log symbolizing the sun's life giving fire was burned. Yellow-sun-fire-gold have been associated from the most remote times.

The ancients knew of nothing that could destroy the transcendent metal. The early alchemists, however, discovered that a mixture of nitric and hydrochloride acids would slowly dissolve the noble metal. They dubbed the powerful solution aqua regia or kingly water. Gold is also affected by a solution of cyanide into which zinc dust is introduced. This discovery, made in the late

19th century, revolutionized gold refining. Immediately used in the gold fields of South Africa, the cyanide process made possible the efficient extraction of gold from otherwise unyielding ore. Combined with the older, less efficient mercury amalgam method the cyanide process recovers as much as 95% of the gold contained in refractory ores as opposed to an unprofitable 60-75% previously extracted. Gold is also soluble in hot selenic or telluric acid.

Gold is yellow when solid and turns chocolate brown when reduced to fine dust. As it boils is gives off a peculiar greenish-purple vapor—a fact that must have enhanced the reputation of goldsmiths as magicians. When concentrated ammonia reacts with man-made gold oxide, an explosive compound may form. This powerful explosive, which is a black powder when dried, is called fulminating gold and may have accounted for many of the mysterious blasts that frequently tore apart the laboratories of alchemical researchers who were the world's first chemists. Another interesting characteristic of gold is its ability to exist in a sol, which is a colloidal dispersion in a liquid. Gold sols may be purple, red or blue depending upon the particle size of the gold. Formaldehyde, tannin and phenylhydrazine are among the common reducing agents added to solutions of gold compounds to generate sols.

The rich texture and beautiful sheen of gold disguise its great density and weight. Gold is extremely heavy. This weight of a gold coin or ornament, satisfyingly heavy in the hand, is part of its appeal. Gold has a specific gravity of 19.32 meaning it is almost 20 times heavier than an equal volume of water. This makes it one and a half times as heavy as lead and twice as heavy as silver. Only platinum, with a specific gravity of 21.45 is heavier. The metal is so compact that a cubic foot of gold weighs about a thousand pounds.

Gold is the most ductile of metals and can be stretched or beaten thin without breaking. A single ounce the size of a sugar lump can be drawn into a fine wire 50 miles long. The same ounce of gold can plate a thousand miles of copper or silver wire. One grain of gold, which is equal to 0.0648 gram, can be beaten to a thinness of 1/250,000th of an inch, producing a sheet about six by six inches. The gossamer gold sheet is translucent, transmits a greenish light and is more than a thousand times thinner than a sheet of paper. A single Troy ounce beaten into a film covers more than 100 square feet. Prehistoric man was quick to discover the malleability of gold, which is so soft in its native state that it can be scratched with a sharp fingernail or stick. Gold reaches the melting point at 1063 degrees centigrade, which was well within the reach of early cooking fires. As gold heats, or even when frozen in liquid oxygen, it doesn't expand, shrink or otherwise distort.

Native gold is chemically snobbish and immiscible, reluctant to form compounds with other elements. It is usually found in relatively pure form in alluvial or placer deposits. It is found in association with most copper and lead deposits and recovered as a by-product and in lodes or veins usually associated with quartz. Gold from placer deposits, lodes or veins is generally in the native or free state, uncombined chemically with other elements. The gold telluride's found in Colorado and eastern Australia are exceptions. In the United States, which ranks a low fourth in world gold production, about a third of each year's gold is recovered from base metal operations. About half comes from the direct treatment of gold ore in which the precious metal is generally invisible, and about 15% comes from placer mining.

Gold, even in its native state, is often combined with a small amount of copper or silver. One such naturally occurring alloy is electrum which is about 20% silver and was highly prized by the ancients. Pure gold is a rich bright yellow. Electrum is the color of a pale winter sun. The tyrant Gyges of Lydia in Asia Minor used electrum to make the world's first coins in the 7th century B.C. Electrum gathered from the sands of the river Pactolus, which flowed down from Mount Tmolus in the highlands of Anatolia, was the basis of Lydia's colossal wealth. The name of one Lydian king, Croesus, still stands for exceptional riches.

Drinking goblets made of electrum were treasured above all else by many an ancient ruler for their apparently magical ability to warn of poison in the wine. Pliny, the Roman historian, wrote that when poison was mixed in such a cup it made a spitting noise and rainbow colors danced in the bowl. Pliny's story may sound incredible but it's true. When wine containing the common poison arsenic was poured into a cup of electrum it did hiss and a rainbow film formed on the surface of the metal. This happened because of the reaction when trace amounts of iron impurities in the metal came into contact with the arsenic forming a film of iron-oxide. The hissing occurred as trapped sulphides and arsenates escaped from the electrum in the form of gas.

At first, many thousands of years ago, gold was worked as it was found. Then smiths learned to combine it with other metals to form alloys. Pure gold is too soft and pliable for most applications. Mixing gold's atoms with those of other metals hardens and strengthens it. It also changes its hue. By varying the proportions it is possible for goldsmiths to create gold pieces in colors ranging from purple to white. Copper is generally added for reddish shades and silver for white. The best whitener is nickel. Joining cadmium and silver with gold results in a greenish alloy. Adding iron to gold turns it blue. A rare alloy of gold and bismuth is blackish in appearance. In Tang China from the 7th to the 9th centuries A.D. goldsmiths perfected a secret process for alloying gold with iron which turned purple when heated. Gold is also combined with precious palladium and platinum for jewelry and industrial use. A gold-platinum alloy was produced in pre-Columbian Ecuador before the rest of the world had discovered platinum. Platinum requires extremely high temperatures before it can be refined. The pre-Columbian goldsmiths couldn't produce such intense heat but formed the alloy by *sintering*—in which a bonded mass of gold and platinum particles were partially fused by the pressure of repeated hammering and heating below the boiling point.

No one is sure how long before 2500 B.C. gold was worked. But by the mid-third millennium, Sumerian goldsmiths were already masters of most of the technical processes used in the history of gold working. The breathtaking beauty of golden objects unearthed from the famed Royal Cemetery at Ur is matched by the high level of their execution. The Sumerians had long ago discovered how to anneal or strengthen and temper gold by heating, cooling and beating it over and over. The Ur artifacts include pieces representing such difficult techniques as raising a complex hollow helmet from a single mass of gold and lost-wax casting.

The purity of gold is referred to as its fineness; the percent of purity or parts of gold per 1,000. Pure gold is called 24 carat. Gold is by far the world's favored metal for jewelry because of its beauty, which can be enhanced by shaping elaborate forms, engraving, enameling or inlaying the surface with stones, other metals, glass or shell, and because of the concentrated wealth it represents.

Purity and color preferences vary widely, influenced by tradition, fashion and even the motive for the purchase. The lower the gold content, the wider the possible range of hues. In parts of Europe gold with a whitish cast is popular especially in pieces set with precious stones. In the United States, where there is no tradition of hoarding gold as a buffer against the ravages of war or inflation, the bulk of jewelry is less pure than in other areas.

In India, Southeast Asia, the Middle East, Africa, and most of Latin America, gold is considered as a form of savings, a hedge against uncertainty and a "special" metal with exalted status. Gold enjoys a 4,000-year-old religious regard in India where it is considered the best form of life insurance by the masses. Indians have a taste for bright deep yellow gold of high purity. But even within the country preferences vary. In West Bengal women who collect gold bangles are satisfied with 22-carat gold which is 22/24ths pure. But in Bihar and Crissa 23- or even 24-carat gold is the traditional favorite. Women are accustomed to having their soft ornaments sag and lose their shape. When this happens, a woman takes her pieces to a goldsmith who transforms them into something new, working in precisely the same way his ancestors did millennia ago.

Indira Ghandi's government, alarmed at the colossal amount of gold immobilized in hoards and at the illegal flow of Indian currency to smugglers who supply the Indians' craving for gold, restricted hoarding and prohibited manufacture of ornaments in more than 18-carat gold. The women of India rebelled and at least a hundred of the country's 450,000 goldsmiths committed suicide while the rest complained they couldn't work the less beautiful and harder metal into traditional Hindu styles.

Americans, who buy gold jewelry for ornament rather than investment, would consider 18-carat gold, which is 75% pure, acceptable for fine jewelry. Very little jewelry is manufactured with a higher gold content and most is marked 10-carat which is only 41.7 pure. By law 10-carat gold jewelry which is used for the mass production of class rings, emblems, club pins and charms can actually be as little as 9.5 carats. The volume of such production ironically makes the United States, which treasures gold the least, the largest consumer of gold in the world.

The English word "carat," which gives its name to the scale used for measuring gold purity, comes from the Arabic "qirat," the Greek "keration," and the Italian "carato," all of which refer to the fruit of the carob tree. The carob, which grew widely in the Ancient World, has long horn-shaped pods. In antiquity the dried seeds, which are of an amazingly uniform weight, were used to balance the scales weighing pearls and gems in oriental bazaars. Each seed weighs about a fifth of a gram which is the weight now assigned to the metric carat, the unit used in weighing diamonds and other gems.

Gold's weight is measured in troy ounces rather than on the familiar avoirdupois scale. The troy ounce takes its name from the French town of Troyes where it originated and equals 1.097 avoirdupois ounces. The avoirdupois system is used to weigh everything but precious metals, gems and drugs. Its name is also of French origin from "avoir de pois" meaning goods of weight but both systems are British.

The systems are based on the most ancient unit of measure —the grain. The troy ounce contains 480 grains, the avoirdupois contains 437.5 grains. Grains were adopted as a unit of weight in ancient Mesopotamia when developing commerce needed standardized weights to make transactions easier. Merchants and traders observed that grains of barley or wheat from the middle of an ear tended to be of standard weight despite the size of the ear itself. They divided the large unit of weight, the shekel, into grains for weighing precious metals. The grain was the hypothetical weight of a wheat grain. In practice a shekel might vary from 120 grains to more than 200. Multiples of the shekel were introduced as trade expanded. The mina was reckoned at from 25 to 60 shekels depending on the time and place, and later the talent was introduced which was equal to 60 minas. The mina and talent were used throughout the Middle East for thousands of years.

The grain was used in Britain until that country adopted the metric system upon joining the Common Market. Today almost every country weighs gold by the gram in actual practice rather than by the grain or archaic pennyweight, although the price of gold on world markets is quoted in troy ounces.

Gold has been put to a remarkable number of uses. The divine and royal metal has been used to awe worshippers, dazzle subjects and seduce lovers and consumers. People instinctively respond to the gleam of gold. In every stage of social evolution gold has had a deep hold on mankind. For as long as men have been plotting against one another gold has been used in espionage. Egyptian and Greek leaders bought intelligence and victory with gold. It still figures in the payment of 20th century spies.

Gold represents anonymous, compact, portable wealth that can be transferred without red tape. It can be divided without losing value. Gold can be melted and poured into any desired shape—a car fender, a shoe heel—which makes it attractive to smugglers. Gold doesn't deteriorate and is so concentrated that a practiced smuggler fitted with a special vest can conceal a small fortune in gold bars under his jacket. Gold quietly crosses borders defying restrictions on its movement. Some $300,000,000 in gold is believed to be smuggled into India each year.

Gold has an unmatched record of almost 4,000 years of service as a monetary metal. Around 2000 B.C. the early Egyptian pharaoh Menes introduced the world to the gold standard. He had standardized 14-gram gold ingots cast more than 1,500 years before gold coins began to replace barter. But before coinage, gold bars and gold rings were used as currency in such far flung areas as Ireland and India. More than 150 substances including whale baleen and butter have been used as money, but gold enjoys unique status as the universally acceptable medium of exchange. It has always been the ultimate store of value, and remains the "payment of last resort" between individuals and nations whatever their differing philosophies and politics.

Gold is being phased out as a monetary metal. Many economists and politicians agree with John Maynard Keynes, the English economist who called gold a "barbarous relic." However, gold has its zealous supporters like the French and Russians who advocate a continued monetary role for the precious metal. Certainly no currency has ever maintained a stability equal to that of gold, although a few, like the post-World War II United State dollar, were as "good as gold" because the paper certificates were backed by actual gold on reserve.

In times of uncertainty gold holds its value. It represents the preservation of capital in the face of inflation, government depreciation, or the wiping out of a complete currency such as that of South Vietnam following the United States' withdrawal in 1975. The Vietnamese, whose history has always been a troubled one, found their piastres worthless, but the little taels of gold they had cached away had actually appreciated in value as the price of gold skyrocketed in 1975 before tapering off. The French, whose currency has been devalued more than 20 times since World War I, trust gold far more than the euro. An estimated 5,000 tons of gold are socked away in French vaults and hiding holes making France second only to India as a nation of hoarders.

Since the end of the 16th Century, when American gold weighted down the galleons that plied the Spanish Main bringing a massive infusion of wealth to weary Europe, fairly accurate records of gold production have been kept. They indicate that an amount somewhat in excess of 120,000 metric tons of gold have been refined. Yet, this mass represents the weight of metal that the U.S. steel industry at its heighth could have poured in less than four hours. Roughly half the gold produced since Columbus discovered America is in currency or bullion. The largest hoard in one place is the almost 15,000 tons of coins and bars stashed away in the impregnable subterranean vaults of the Federal Reserve Bank in New York City. The rest has been utilized in arts and industry or is in private hoards.

Artistic uses of gold vary. Gold jewelry, amulets and idols were the earliest products of the goldsmiths. Soon craftsmen learned to embellish every conceivable surface including shell, clay, stone, paper, wood, leather, fabric, embroidered cloths, other metals; then, glass, and, more recently, plastic and other synthetics.

Gold is an ideal coating material furnishing beauty, insulation and superb protection against the ravages of time and the elements. Contemporary architects use gold coatings manufactured at relatively low cost in building churches, public buildings, and hotels and office buildings as well. The gilded domes and spires of temples, churches and capitol buildings proclaim their spiritual ties or their importance. In other types of buildings gold-plated tiles reflect heat and glare. Gold-coated facades or roofs are impervious to the cancer of chemicals which eat away stone, and to the rot that destroys wood. From Australia to New York large buildings with great expanses of glass rely on finely spun gold imbedded in the glass to cut down on heat and light along with air conditioning costs. The world takes on a greenish tinge when viewed from behind the golden walls or the gold film, less than two millionths of an inch thick, which is sandwiched between layers of glass to make the heated windshield for some aircraft. Golden windshields prevent misting and icing. They also reflect the glare of the sun and filter out its harmful rays.

The capacity of gold to reflect up to 98% of incident infrared radiation and ultra-violet rays makes it a valuable material for space apparatus and instrumentation. Edward White, the first American to walk in space used a gold coating on the umbilical cord which connected him to the Gemini spacecraft. Gold was used on the moon buggies, the astronauts' helmet visors and many other pieces of equipment. It is such a splendid reflector of heat that a gossamer film of gold on the heat shield of a rocket engine can protect the fragile instruments inside from the searing heat generated by lift-off thrust and the re-entry into Earth's atmosphere. Commercial aircraft have gold-plated engines to control engine temperatures.

The first human artifact that left our solar system was gold. In 1972 NASA launched Pioneer 10 on a journey that will carry it far into the cosmos. A small gold-covered plaque bearing outline drawings of a nude man and woman with explanatory scientific symbols was attached to the antenna support strut. It won't reach the first star on its route until 80,000 years from now. But when it does, and for millions of years after, it should be lustrous and beautiful, eternally free of tarnish, erosion or scale.

It was fitting that mankind's first message to worlds beyond was made of gold, the sun's own metal. But it isn't gold's beauty or link with the sun that have recommended it to modern technology. Gold is used in consumer, defense and space industries because of its resistance to change, its high conductivity of electricity and its superior reflective and insulating qualities.

Modern electronics and computers take advantage of the metal's excellent conduction of electricity – exceeded only by silver or copper, both of which are subject to deterioration. A miniscule amount of gold in a colloidal suspension on a printed circuit can actually replace miles of wiring. Wherever absolute dependability is essential, gold is used. The sensitive circuitry of the computers that guided men to the moon was covered with gold. Gold performs perfectly in the circuitry of repeater units in the communications cables that lie on the ocean bottom. It is in the minute circuitry of transistors, telecommunications equipment, radar, televisions and computers. Gold plating as thin as three- or four-millionths of an inch is enough to protect components from corrosion.

Modern technology can produce a translucent golden film large enough to cover an acre and fine enough so that it would take about a thousand such sheets to equal the thickness of this page from a mere ounce of gold. But even in antiquity, gold foil was made and widely used. Goldbeating is one of the earliest crafts. The gold leaf found on the mummy of the pharaoh An-Antef, who died four and half thousand years ago, was made in much the same way it is still prepared today.

The Egyptians applied gold leaf generously to furniture, royal barges and statues. The Chinese used gold foil to ornament wood, pottery and textiles. The ancient Greeks learned the arts of goldbeating and gilding from Egypt, and in turn passed on their knowledge to the Romans. Gold dust mixed with chalk, powdered marble and glue was applied to large and small surfaces. The Greeks not only gilded sculpture, masonry and wood, but also fire-gilded metal by applying a gold mercury amalgam to a metal surface and then driving off the mercury with heat. The poisonous mercury accounted for the loss of hair, tremors and blurred vision that afflicted many gilders. In spite of the dangers of fire-gilding it was practiced until fairly recent times and caused many deaths. An 18th century A.D. chronicle describes fire-gilding in London and notes that the majority of those involved were women.

Ancient Rome was a dazzling city. Gold leaf was lavished on the facades and interiors of palaces, temples and public buildings. The rich flaunted their fortunes favoring showy displays over elegant refinement. They used gold for everything—to cover beds, to gild the horns of sacrificial animals, for jewelry, plate, statues and medicine. According to Pliny, gold beating in ancient Rome was a highly developed art. In their time, in the first century A.D., skilled beaters could turn an ounce of gold into 54 square feet of gold sheet in the form of as many as 750 tissue thin leaves. Each leaf measured about three inches square. By the mid-1600s European goldbeaters had advanced the exacting art so that one ounce of gold could be beaten into leaves which together covered an area of 105 square feet.

In London, where there was an organized guild of gold beaters from the 12th century A.D., gold leaf is still produced by the ancient technique. An apprentice spends many years to become a first class gold beater. He learns how to hammer an ounce of pure gold between sheets of vellum and then as it thins, between specially treated membranes from the intestine of an ox. It takes a series of graduated hammers and as many as 400 of the extraordinarily thin but highly resilient ox membranes to treat an ounce of the precious metal and turn it into 800 leaves of gold film an astounding three millionths of an inch thick. This is an expensive business. Each membrane must be replaced after a thousand hours making the value of each equal to that of the square of gold film laid on it.

It is the hand labor however, that makes traditional gold leaf so costly. In Asia, where wages are low, gold leaf costs far less. Worshippers can show their devotion by offering small squares of gold leaf to gild temple architecture and statuary. In the West a less expensive way of producing filmy gold has been developed. It is used for goods not gods as a glamorizing film to entice consumers into purchasing glassware, the contents of bottles and jars, china, greeting cards, books, and hundreds of other products. Liquid gold, as little as four millionths of an inch thick, is stamped or sprayed on everything from whiskey bottles to shirts.

One of the oldest uses of gold was in medicine. The inalterability of the sun's metal and its association with the divine led the ancients to regard gold as the universal panacea linked with the idea of eternal life. The Chinese, whose word "chin" meant both gold and metal, believed gold leaf was the most perfect form of matter. Chinese alchemists and doctors prescribed gold-based medicines to give renewed life to ailing bodies and prolonged life to healthy ones.

The Rig Veda, the most ancient of the four sacred books of Hinduism, says the giver of gold will have a life of light and glory. Hindus believe that giving away a bit of gold cleanses ones sins. A devout Hindu often carries a bit of gold with him ready to give away. A father touches his newborn child with gold to bring good fortune and prosperity. At death a bit of gold is put into the mouth or among the clothing of a devout Hindu. Gold leaf is eaten at celebrations for good luck and doctors prescribe gold to cure infections and treat a host of other ills.

Archaeologists recently excavated an Egyptian tomb and discovered the first known instance of dental application of gold. Some 4,500 years ago a man had a gold bridge made. No doubt the ancient dentist used gold for the same reasons of a modern dentist. Today gold is used for caps, bridges, fillings and inlays not for its mystical sun link but because it is malleable, acid-resistant, tasteless and non-poisonous.

In the West gold has a long history of medicinal use with bright prospects for new applications. The most celebrated medicine of the Middle Ages was aurum potable, the elixir of gold dissolved in liquid, which was prescribed for everything from fever to plague. It is first mentioned in the Bible. When Moses returned from his mountain top conference with God he found his people had turned from the Lord and returned to worship of a golden idol. Moses was so enraged that he had the blasphemous golden calf, which had been made from donations of gold jewelry, ground to a powder. He mixed the dust with water and forced his people to drink it.

In ancient Rome gold-based ointments were ordered for skin ulcers, ringworm, fistulas and hemorrhoids. In Medieval times herbalists gathered medicinal plants with golden sickles just as the Druids before them had cut the mistletoe, the "sacred vegetable gold" which they thought came from the sun, and used in their ceremonies at places like Stonehenge in England. Salts, elixirs, unguents, pills and injections of the eternal metal have always been popular. In the early 20th century, gold injections were used for the cure of alcoholism and syphilis.

Injections of gold salts are used to treat some forms of arthritis in chrysotherapy (from the Greek *chrysos* meaning gold). Gold dissolved in an oily mustard-like liquid is injected over a prolonged period to help alleviate arthritic pain. Radioactive gold has been used since the 1940s to irradiate cancerous tumors. Gold-silver alloys have been used to repair skull injuries, and Western medicine still relies on gold as an ingredient in the treatment of certain nerve-end conditions, burns and ulcers.

Gold has also been used by charlatans and the superstitious. Astrological healers claimed that gold medicines were even more effective taken under certain of the sun signs. Gold leaf inscribed with magic symbols was taken for pneumonia, tuberculosis and appendicitis. An American folk remedy for a sty is to rub the afflicted eye nine times with a gold wedding band. In 17th-century England wealthy ladies who had been ravaged by smallpox applied an ointment of gold, salt, ceruse, rosewater, camphor, glycerin and vinegar to their scars which was supposed to "raise up and fill the hollows and pits."

The most powerful therapeutic effect of gold was mentioned by the Italian Biringuccio in his 16th-century treatise on metallurgy. "Indeed as a medicine it is beneficial to certain illnesses. Nature with her own virtue has endowed gold, as a singular privilege, with power to comfort weakness of the heart and to introduce there joy and happiness, disposing the heart to magnanimity and generosity of works. Many learned men say that this power has been conceded to it by the benign influence of the sun and that for this reason it gives so much pleasure and benefit with its great powers; especially," he added, "to those who have great sacks and chests full of it."

The belief that gold came from the sun was still strong in the 16th century. Every metal was thought to "receive a special influence from its own particular planet," according to Calbus of Freiberg, who wrote that "Gold is of the Sun or its influence, silver of the moon, tin of Jupiter...Thus gold is often called the Sun."

The ancients were on the right track. The lustrous metal they held sacred as the seed or offspring of the sun is indeed "sun stuff." But it didn't come from our young sun which is a cyclotron, producing energy as hydrogen turns into helium providing the heat and light which sustain life on Earth. Gold in the form of searing gas is present in our sun. There may be as much as fifty billion, trillion troy ounces of it, which may someday go to enrich worlds not yet born.

All of the gold distributed in the crust of our planet came from dying suns at the fiery birth of Earth some four billion years ago. A star is generated out of dust and interstellar gas composed of hydrogen atoms. The stellar core heats up and nuclear reactions, similar to those in a hydrogen bomb, occur, which are part of the star's life cycle. The abundant hydrogen atoms are burned to form helium — the two simplest and lightest elements. As a star evolves it consumes all the hydrogen at its core and must burn helium. Finally it burns carbon and when that too is exhausted the stellar core turns to pure iron which is inert and can't serve as fuel under usual conditions.

Lighter stars then cool off and die. Something different happens to massive stars. They are less stable and undergo cataclysmic explosions. These red-hot bursting stars are called supernovae. In a final great death surge they become as brilliant as the whole galaxy and generate masses of heavy atoms including gold. Our entire solar system was formed out of a gold-enriched interstellar medium which astrophysicists call the solar nebula. How did such fiery stardust enter the molten mass of magma at the infant Earth's core and end up in the Egyptian deserts, the California mountains, or the South African goldfields?

The heavy elements, including gold, silver and uranium, were hurled out into space at a speed of 3,000 miles a second. They floated about until a gassy cloud formed that swirled around our sun. Eventually the cloud collapsed under local gravitational condensation. The heavier atoms precipitated out first to form the dense rock and metal laden planets closest to the sun—Mercury, Venus, Earth, Mars and the asteroids that orbit between Mars and Jupiter. By the time the more distant planets formed there was little left of the heavy elements which had condensed on the smaller planets. As the Earth cooled, the gold, because of its extreme weight, settled beneath the mantle. Only 18 miles below the thin solidified crust of the Earth is its molten core. This mass of magma has been boiling for eons. Convulsive movements stretched and snapped the Earth's crust. Mountains and volcanoes were thrust upward and the mantle was twisted, wrinkled and fractured. Molten magma from the Earth's liquid core has been forced upward at incredible pressure through the crust into cooler zones of sub-surface rock.

The metallic elements were carried in suspension, and as the magma cooled, formed masses of rock, often granite, the base of mountain ranges. The metallic ores—including gold, silver, platinum, tungsten and copper, were deposited in fissures and cracks of faulted rock. They took the form of metallic oxides or, in the case of gold, of the solid mineral.

It is one of Nature's jests that gold, so precious and rare, is found almost everywhere on Earth. Gold is present in polar ice, human hair, plants and deer antlers, for example, but in minute amounts. Philadelphia is built over an untapped goldfield and many of the city's buildings are made of bricks in which the gold is invisible. In Brazil the itambamba plant and in the Philippines the gogo plant yield gold. Certain gasses in Canada have been burned to recover the microscopic gold they contained.

The greatest potential source of gold is water. The word "treasure" conjures images of sunken galleons filled with chests of bright gleaming gold over which skeletons and monster octopuses, stand guard. Yes, this kind of treasure does exist. There are hundreds of millions of dollars in sunken treasure awaiting men of will and daring. But it all pales beside the colossal amounts of gold suspended in the salt waters covering almost three-quarters of the planet. There is enough gold, tantalizingly beyond reach, in the oceans to make every human being a millionaire...if only a profitable way could be found to extract it from its salty solution.

How did gold get into the more than 130,000,000 square miles of ocean? Some was laid down in placer deposits beneath the ocean floor when Earth was formed and was later thrust up during geologic convulsions that formed undersea mountains. Ever since

Earth's superheated halo condensed to form the oceans, gold has been washing into them. Erosion tore gold from rocks, and rain, ice and wind aided by gravity sent it lower. Streams in gold-bearing areas such as Alaska, Japan, Australia and Oregon deposited rich concentrations in the Pacific. Colloidal gold precipitates to the sea bed or is stored in natural aquatic gold mines such as fish, plankton and seaweed. Traces of gold are present in the nodules which cover about 22,000,000 square miles of ocean floor although their chief components are manganese, copper, nickel and cobalt. Ever since 1866 when a French scientist speaking before the American Association of Science announced: "Gentlemen, I believe that among other things you will find gold in the ocean," scientists have been trying to find a method of harvesting sea gold. The first scientific estimate of the amount present was made by an English chemist in 1887 who arrived at a figure of 65 milligrams of gold for every metric ton of water. Later estimates varied widely but pointed to several interesting facts. There proved to be more gold concentrated in polar ice than in surrounding waters. The highest levels of all were present in plankton, the minute drifting plants and animals fed on by many marine species. Although levels vary from zone to zone, the average concentration is only one to two parts of gold to a million parts of water.

Various filtration and evaporation schemes have been tried and abandoned. A totally efficient process would require more than a billion gallons of water to be processed for every ounce of gold recovered. Some scientists think that nuclear powered desalinization plants in the future might produce gold as a by-product. Meanwhile, mining companies are prospecting for undersea placer deposits where the gold is in solid form.

Until technology can solve the puzzle of how to profitably separate the billions of tons of ocean gold from ocean water, gold production will continue to come from the parts of the Earth where nature was most generous with the precious metal. The greatest and longest-worked deposits are in Africa. South Africa, which a century ago produced less than 1% of the worlds gold, today accounts for over 70% of annual production, excluding the Soviet Union, which doesn't make gold production figures public, but is the world's second largest producer. The parts of Africa that once yielded so much gold, Egypt, Ethiopia, the Sudan, Central and West Africa still furnish small amounts of gold.

Asia Minor, Arabia, the Hindu Xush, southern and eastern India have been important gold sources in the past. Asiatic Russia has vast deposits of gold. Gold in considerable quantities has been mined from the Chinese coast, Japan, Korea, the Philippines and through the Malay Archipelago, Indonesia, Papua New Guinea, Australia and New Zealand.

The gold of Europe was concentrated south-east from the British Isles, through France, along the Alps and Balkans, and in extremely rich deposits in Spain and Portugal. In the Western Hemisphere gold deposits were laid down the western coastal areas from Alaska to the northern half of South America, running across part of the western United States, as far east as Georgia, and across Mexico and much of Central America and some of the Caribbean islands.

Gold is found as dust, flakes, grains, nuggets, in curious crystalline branch-like formations and in ores where it is often invisible to the naked eye. A small amount occurs combined with tellurium which is a rare, brittle, non-metallic element used as a tint for glass and in alloys. The branching formations of gold were offered as evidence for the belief that gold grew as a plant among rocks. European miners until modern times believed gold deposits should periodically be closed to allow the precious metal time to regenerate. Miners in various parts of the West Indies prayed and fasted before gathering gold and were careful not to remove all they saw so the "soul" of the gold wouldn't go and leave the earth forever.

Most gold is found in primary vein deposits or secondary alluvial deposits. Gold veins are so called because they appear to wind through the host rock like the veins of the human body. They are usually associated with a quartz matrix and can be several miles long or end after six inches. A vein or lode can be as narrow as a string at the surface, divide into almost non-existent branches and then plunge deep into a rich pocket of high grade gold. There is no way to predict the course of a vein as many a frustrated prospector has discovered. Prehistoric river beds, long covered over, and the gold bearing-reefs or "bankets" of South Africa's Witwatersrand are examples of gold deposits which originated from quartz lodes or veins.

The world's greatest gold field is the Witwatersrand "reef." The term "reef" was coined by sailors who jumped ship during the Australian gold rush of 1851. It referred to the way the quartz outcroppings projected above the surrounding terrain the way coral reefs rise from tropic seas. The gold of the Witwatersrand is imprisoned in a hard quartz matrix extending deep into the ground. Giant crushing mills and tons of acid and other ingredients are needed to retrieve one ounce of gold for every three tons of processed ore. Colossal amounts of capital and highly sophisticated technology are required for such large scale mining and refining. It costs a hundred million dollars merely to bring a new mine into production. It has never been the miners who got rich hacking invisible gold out of galleries two miles deep in the world's most golden zone.

Quartz is by far the most common matrix for gold veins but they are sometimes found in calcite or limestone formations. If gold was in an active solution during a period of igneous activity and was deposited in rock such as limestone, which is vulnerable to chemical action, it formed what is called a replacement deposit. In this case the gold solution penetrated the solid but soft rock and slowly dissolved it, replacing it, atom for atom, with gold. Gold in replacement deposits is always found associated with gangue, the valueless mineral matter and rock which hold it in place.

Before 1873 when deep vein mining was introduced to work the rich lodes in the American West, alluvial or placer mining accounted for most of the world's gold production. This gold was the product of erosive forces which freed the metal from hills and mountains. Pieces of loosed gold ranging in size from boulders to fine particles were carried into streams and rivers. Through the eons the earth has been worn down by the weathering forces of rain, wind, freezing, thawing, plant growth and gravity. These natural agents uncovered hidden gold as they removed hundreds, even thousands, of feet from mountain ranges such as the Himalayan or California's Sierra Nevada.

Chunks of gold and gold-bearing ore were tumbled down streams as they flowed toward the sea. Bumped continually against rocks, scoured by sand and pebbles, gold nuggets became smooth and pieces of ore lost their waste material. Rock was worn away and surface water carried off some of the constituents of metal bearing minerals, which are subject to chemical decomposition. Silver, for example, decomposes upon contact with chlorides. Rainwater contains chlorides as it swells streams and rivers so that silver is not found in alluvial deposits. But gold, impervious to chemical assault, collected in watercourses and former watercourses through the countless millennia. Shining nuggets and sparkling dust lay waiting for homo sapiens, hunter and gold seeker.

Alluvial deposits exist underground too. In California and Australia there are gold-bearing gravels lying deep beneath the thick flows of ancient basaltic lava. The gold in the fossil placer deposits was carried into prehistoric streambeds at the dawn of time. At a later period geologic processes welded the gold to the bedrock which was eventually covered by sedimentary or igneous rock. These subterranean alluvial deposits have been discovered when they were exposed by later streams which channeled through the ancient layers of overburden.

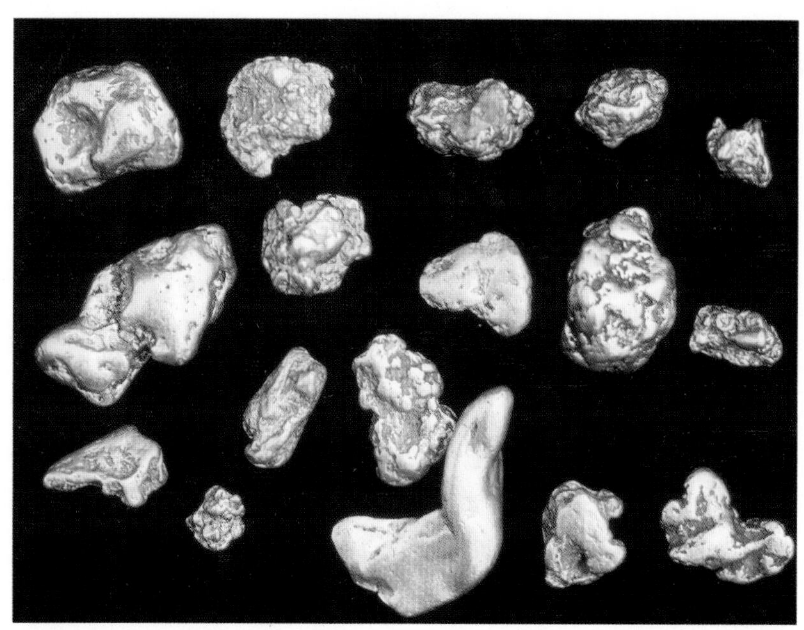

These gold nuggets in their various sizes and shapes are examples of those that can be found freely in nature.

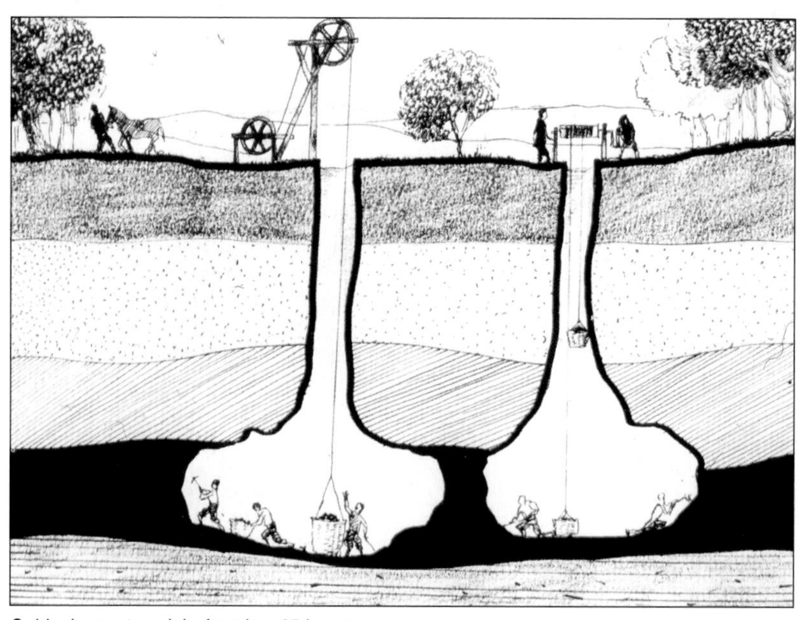

Gold miners at work in Austria—15th century.

(Above, left) Golden present from Czar Nicolas II to his wife Alexandra in 1904.

(Above, right) Ornate Viking goldwork on figurehead on ships tiller, circa 9th century.

5th century Norton's Ordinal Alchemist.

A 16th century Indonesian Kris made of gold with precious stones on handle—recovered from Portuguese wreck off of Sumatra.

(Below, left) Solid gold with precious stones French Reliquary of St. Foy, 11th century.
(Below, right) Renaissance jeweler.

(Left) Early 17th century unidentified Dutch noble flaunting his wealth in the form of heavy gold chains. *(Below)* Sovereign of Henry VIII (gold obverse). King Henry on throne was minted 1544–47.

(Below) British gold noble minted during reign of Edward IV.

Chapter 2

The Golden Age

Once upon a time when the world was fresh and new the gods created a race of men who lived in peace and plenty. They were golden men and theirs was truly a Golden Age for they found the Earth filled with precious metal. Great forest fires kindled by sacred lightning liquefied gold, smelting it out of lofty rocks and cascading it down mountain slopes. Earth was encircled by Ocean, the deep, broad river, which was the parent of all the waters. According to the Classical Greek legend of that long vanished time, golden sands lined Ocean's shores and islands made of shining gold floated on its surface. After a while when the golden beings began to squabble among themselves, the gods replaced them with a race of silver men. As these new inhabitants of Earth became violent and selfish, the angry gods created men of bronze who proved even more warlike and unworthy than their predecessors. Finally, the disappointed immortals peopled the Earth with men of inferior iron, and it was into that Iron Age that the Greek poet, Hesiod, who told the story in the 8th century B.C., lamented he had been born. The Golden Age lingered in legend alone and the common lot of man was to toil without reward.

Gold was indeed scarce in the early days of Classical Greece except in the verses of the poets which were strewn with "golden dawns," "golden warriors," "golden loves" and "golden youths." The great philosopher Plato explained gold's rarity and set off on a treasure hunt, which has lasted more than 2,000 years. He wrote of the continent of Atlantis where mankind once dwelt in golden luxury and harmony with the gods. It lay just beyond the Pillars of Hercules, the modern Strait of Gibraltar, in the awesome Atlantic which was known as "The Green Sea of Darkness".

Each day Apollo drove the chariot of the sun across the heavens, charting his course by the golden spires of Atlantis and the world was a happy place. But the mortals over-reached themselves and bred with the gods producing inferior beings. This made Zeus, the supreme deity, so angry that he cast Atlantis—gold, inhabitants and all—into the fathomless depths of the sea. Since that time men have been condemned to comb the earth and literally to slave for meager bits of the gold which had been so generously lavished on their ancestors. The quest for Atlantis continues. Romantics, mystics and even scientists have searched the seas from the Mediterranean to the Bahamas for a trace of the lost continent whose golden spires once lit up the world.

These Greek myths and others with a golden motif may have at their heart a dim memory of the prehistoric era when the number of men was small and the accumulated golden treasure, which nature had been mining for eons, was large. The story of gold is one of treasure and of time measured, not in years or centuries but in millennia. Billions of years ago elemental forces began mining the heavy yellow metal. Millions of years ago grains and nuggets torn from soaring mountains and carried into streams along with crumbled rock began to collect in the alluvial gravels in the upper reaches of rivers to form placer deposits. Smaller pieces were carried lower before they sank, and some particles, as fine as flour dust, floated right out to sea. When the first man-like beings appeared on earth the treasure of infinite ages already lay waiting for harvest like a crop of gleaming berries.

Who were history's first gold seekers? Where did they find gold and what did they do with it? Organized gold production began with the Egyptians some six or seven thousand years ago, but long before then people found and cherished bits of the gleaming metal. We can only conjecture about that dim prehistoric period when gold, glowing like nothing else in nature save sunlight or fire, must have caught the eye of our earliest ancestors as they roved in search of fruits or followed the natural pathways of streams hunting fish and game. There isn't any archaeological proof that the first humans paid any attention to the gold that spangled river sands and shone bright in limpid streams. But it seems likely that men who had mastered the art of making fire half a million years ago would have picked up a smooth tumbled nugget if they saw one winking in the shallows or caught in the tangled exposed roots of a storm-felled tree. It seems possible that the men in Europe, Western Asia and Africa who produced highly efficient flint hand axes a quarter of a million years ago would have discovered some of the remarkable properties of gold. It may have been no more than a shiny bauble to those primitive toolmakers, or perhaps it was already regarded as a very special substance.

Cro-Magnon men living in Spanish mountain caves over 30,000 years ago evidently knew the gleaming lumps could be easily shaped by hitting them with a stone. The cave dwellers who were the ancestors of today's Europeans, physically not much different than modern man, lived in one of the world's richest storehouses of golden treasure. They made remarkable ritual paintings in some of their caves and archaeologists exploring them have unearthed a few bits of gold rudely hammered into small disks, from a layer of occupation formed during the fourth glacial period.

Paleo-men in other areas knew gold, and stone age burials in Nubia have yielded gold-tipped arrows and a gold handled flint knife. Gold was relatively plentiful in remote times but as long as man was consumed by hunting and gathering activities he had little time and energy for prospecting and production. Thus, the history of gold begins with civilization which emerged in the near past, about 8000 B.C., and ushered in the favorable climate of the Mediterranean world. The knowledge of working gold parallels the rise of the early cultures which were characterized by the development of farming, the use of metals and the birth of true cities with complex social organization.

In the last century and a half the golden legacy of many ancient cultures has dazzled an unsuspecting world, which had no idea of the breathtaking treasures crafted thousands of years ago by Sumerians, Egyptians, Minoans, Trojans and Mycenaeans. Now, museums the world over display the splendid artifacts which bring to life the history of the people who lived and laughed and died so long ago. The tangible evidence of the creative brilliance and technical skill of prehistoric goldsmiths in the Near East and Mediterranean is overwhelming. It has eclipsed the role gold played among less spectacular cultures.

What of the metal's earliest associations and of the role of gold in prehistoric Europe? Tracing the glittering thread takes us on a fascinating journey through time and space from the dawn of civilization to the battlefields where gold-clad Celtic chieftains vainly fought Roman invaders.

About 8000 B.C. the "invention" of agriculture, which anthropologists refer to as the Neolithic evolution, made man the dominant animal on earth. After more than a million years of roaming in small foraging bands like other animals, hunter-gatherers living on the hilly flanks of mountains in the Near East began to act upon their environment rather than reacting to it. The development of primitive agriculture appears to have taken place independently and at different times in or near the productive strip of land which curves around the Arabian Desert from Iran, through Iraq and Syria and down the Eastern Mediterranean to the Nile Valley. Within three or four thousand years agricultural communities had emerged in such widely scattered places as Peru, Mexico, Northern China and probably parts of Southeast Asia.

Agriculture was the basis of civilization. Instead of depending on the bounty of nature, the first farmers learned to sow seed harvested from wild barley and wheat. The planting of seed saved from a previous harvest, cultivating, reaping and the domestication and breeding of animals that had roamed wild all had a profound effect on mankind. When small bands of proto-farmers settled in permanent communities and could depend on a fairly regular food supply, they had the energy and time to become civilized.

With farming came surplus food which made possible a dramatic increase in population with the consequent elaboration of the way people lived. Increasingly complex communities stimulated the evolution of government, public works, technology, communication and commerce as well as the elaboration of religion and mysticisms.

People were able to plan for surplus production, which could be bartered within the community and with other areas for desired raw materials and "luxury" items. Among the first trade items were salt, amber, obsidian, red iron ore used as pigment and, of course, gold.

The first Near Eastern cities were not just overgrown villages, but real urban centers where life was varied. The division of labor created social classes with differing levels of power and responsibility. From the beginning, an elite existed which claimed authority and privilege. The ruling class was priestly and the possession of gold was restricted to them and the all-powerful gods they embodied or represented. The most precious substance reinforced their aura of divinity. Gold had been linked with the sun ever since men comprehended the crucial role the sun played in agriculture. A man holding piece of glimmering yellow metal felt a mystic tie to the heavenly fire which sustained life.

Sunlight and gold were important elements in the creation myths of many ancient cultures. In antiquity people used myths to explain the past, unravel the mysteries of nature and account for present conditions. Gold, perhaps more than any other substance, was the principle ingredient in the mythologies of widely scattered peoples. The complex Icelandic and Scandinavian sagas have gold as the dominant motif. In the *Elder Edda*—which has its roots in pre-literate Norse myths—and in the great Teutonic epic, the *Hibelungenlied,* gold lust sets events in motion and becomes a curse in the form of a ring which brings tragedy to its successive owners.

Gold was credited with the power of creation. Not only did it increase itself but it could beget offspring as in an early Siberian legend and the familiar Greek myth of Danae. Acrisius, the king of Argos, was told by an oracle that his beautiful young daughter would one day bear a son who would kill him and seize his throne. To protect himself the king shut Danae up in an impregnable tower of bronze and decreed she should spend her entire life there. But the great Zeus looked down from Olympus, home of the gods, and seeing how lovely the tender princess was, he fell deeply in love with her. Even the strongest prison couldn't keep the mighty god from Danae. He changed himself into a radiant shower of gold, so the story goes, and entered the bronze tower through the bars on the window. As result she gave birth to the hero Perseus, who did eventually kill his grandfather and become king as the oracle had foretold.

The Scythians, who were among the greatest gold lovers in history, had a myth of sacred gold which fell from heaven, and cultures from Africa to North American Indians thought gold had fallen from the sun. Until fairly recently gypsies and some people in the remote parts of Central and Eastern Europe considered gold the "seed of heaven." The parasitic mistletoe which flourishes without roots in trees in many parts of Europe was ritually linked with gold and the sun for thousands of years. The glow of the waxy berries in the bare wintry branches of trees gave rise to the belief it is the earthly form of the heavenly fire god, the sun.

Mistletoe, the "golden bough" of Virgil and the golden fern seed described by Sir James Fraser in his famous work on mythology *The Golden Bough* played a prominent part in legends and ceremonies involving fire, sun and gold in Great Britain, Spain, Scandinavia, Russia, Switzerland, the Italian Tyrol and Brittany. The most familiar ceremonies were those of the Druids who gathered mistletoe with golden sickles at prescribed times in the year. Mistletoe is still used as a remedy in European folk medicine.

Throughout Europe and northwestern Asia great ceremonies linking sun-life-gold were held on Midsummer's Eve. They marked the summer solstice when the sun was farthest from the equator. As the following days grew progressively shorter men kindled huge bonfires in a vain effort to give renewed strength and fire to the sun and thus prolong the daylight which nourished their crops.

Whoever was lucky enough to collect a piece of "the golden bough" or the magic golden "fern seed" on that special night believed he would find actual gold. In Sweden a twig of the vegetable gold was used as a divining rod which was said to dip towards the earth in response to buried gold. In some places finding gold on Midsummer's Eve was as simple as tossing a twig of mistletoe in the air and digging where it landed. But finding gold has seldom been as simple as that, not even for the prehistoric metal workers collected from the harvest of many ages.

In the earliest civilized communities a select group of artisans became metalworkers. They were the men who had been most skilled at collecting suitable stones and working them into tools, weapons, cult objects and ornaments. These specialists were unusual men who practiced a craft that demanded their full attention and were accorded high status in the community. People regarded the metal worker with awe because of his apparently magical ability to transform matter. He could take a lump of gold and turn it into filmy sheet. In the goldsmith's strong but sensitive hands the shimmering yellow expanse then became golden flower or cowry shell or perhaps a delicate bowl whose surface he embellished with raised or incised patterns. Just as amazing, he could take a dull, rough chunk of copper ore and turn it into glistening mass of molten metal which he cast into tools or weapons.

Neolithic people in the Near East knew three metals—gold, copper and extremely rare meteoric iron. Experts disagree about which came first. Gold and copper were widely distributed in some areas of the Fertile Crescent but not found in others such a Mesopotamia between the Tigris and Euphrates rivers. Gold, copper and meteoric iron are the only metals that occur in nearly pure metallic form and display significant luster in their native state. The surface of copper and of iron however can be dulled by oxidation, while nothing can hide the extraordinary sheen of a gold nugget. Cyril Stanley Smith, the noted metallurgical historian writes that "Metals appeared because millennia ago someone's artistic sensibilities were piqued by an interesting and pretty stone." What could be prettier than bit of gold? It seems probable that gold, by virtue of its greater color and luster, first caught man's eye and tempted his hand with its superior malleability.

Iron, next to aluminum, is Earth's most abundant metal. But it wasn't until ca. 2700 B.C. that primitive smiths could produce the intense heat needed to smelt iron from its ores. It took other 1,500 years to develop the relatively sophisticated techniques of iron-working which made it possible for Near Eastern people who couldn't afford expensive bronze implements to replace their stone tools and weapons with iron. However, iron in the form of siderites, nearly pure meteorite chunks of the metal, were known to early men. The Sumerians recognized its origin and called it "Heaven's Metal." Chunks of meteoric iron ranging in size from pebbles to masses weighing hundreds of tons fell to earth in random locations. We still don't know how smiths managed to work the hard metal into iron beads and an iron knife found in two 6,000-year-old sites in Egypt.

The earliest experimentation with materials was prompted by artistic motivation. Long before the smelting of copper ores, people were grinding up colorful rocks to make cosmetics and paint. Traces of eye shadow and rouge from pulverized rock have been found in ceramic pots on many early sites. Some 8,000 years ago artists in the Anatolian town of Catal Huyuk, one of prehistory's largest and most fascinating communities, executed a remarkable series of murals on the plastered walls of religious shrines. Archeologists excavating the site found the colors still vivid—blues and greens from powdered copper ores probably mixed with animal fat, purple from manganese ore, gray from galena, the ore yielding lead and silver, black from soot and reds, yellows and browns derived from iron oxides.

At some point an observant individual noticed that among the green copper ore he was grinding there were bits of shiny reddish material that didn't break when he pounded them. Copper ores occur in much of the Near East and copper, which is soft enough to be worked with a store hammer but hard enough to take a fair edge, was without doubt mankind's first utilitarian metal. The oldest metal artifact from the Near East is not a tool but an inch long copper pendant made about 9500 B.C. The oval ornament was found in a cave of the Zagros Mountains of northeastern Iraq.

The first examples of copper tools date much later. Archaeologists excavating the embryonic farming community of Cayonu in southern Turkey found a few copper tools—a primitive gouge and three sharp pins, one bent like a hook—dating back to about 7200 B.C. Nearby is a copper deposit which is still being worked, 9,000 years later!

Archaeology has provided evidence that by the 6th millennium B.C. the technique of copper working had begun to spread throughout the Near East. Scientists have found a variety of early copper artifacts at sites far from copper deposits, indicating the metal was traded among early people. Gold, too, must have found its way into areas where it wasn't native—most likely reserved for gods and rulers. Unfortunately, none of the golden artifacts crafted by the first metalworkers have been found. But any day an archaeologist may unearth a golden confection which would clarify the appearance of worked gold in the first civilizations.

Both gold and copper could be worked by men with limited skills. Gold offered some advantages. Copper hardens as it is hammered. Gold doesn't and can be beaten into fairly thin sheets with a stone hammer on a stone anvil without having to be heated periodically to prevent cracking. Gold, unlike copper, can be welded simply by hammering pieces together. Copper, however, has the advantage of being hard enough to be useful for implements, although such simple golden tools as fish hooks like those Columbus saw West Indian natives using in 1492 are being used today in gold-rich areas like parts of Colombia.

Little by little, toiling in secret and guarding their knowledge, the metal smiths puzzled out the techniques of working gold, copper and silver which they passed on to the next generation. Through countless years of lonely experimentation, sweat and fire, they learned to extract metals from their ores and create useful alloys. They perfected the smelting of bronze, the efficient alloy of copper and tin (at first they used arsenic with the copper), and then much later, the smelting of iron, the metal which greatly accelerated the pace of human progress.

By 4000 B.C. Near Eastern metal smiths were practicing a crude form of copper smelting and no longer had to depend on the availability of chunks of native copper. They could also extract silver and lead from their ores by smelting them in a charcoal fire burning in an enclosed space to produce intense heat. Copper mining and refining represented one of the world's first industries.

Two of the earliest centers of copper production on an industrial scale have been found and excavated by archaeologists. One is at Tal-I-Iblis, today a barren rock and sand-strewn site in Iran, where copper was produced by the beginning of the 5th millennium. So many crucibles and other artifacts were found that some archeologists speculate enough copper was produced to supply markets as distant as the cities of Sumer, which were springing up along the silted banks of the Tigris and Euphrates rivers. About the same time, the copper industry in the Tima Valley of Israel began on a smaller scale. As the demand for copper increased with the invention of bronze, foreign miners, particularly Egyptians and their slaves, swelled the numbers involved in what became a complex commercial enterprise supplying metal to the pharaohs by the 2nd millennium.

By the third millennium, tin was being used in the manufacture of bronze, which was more versatile and durable than copper. The advent of such metallurgy gave its name to the Bronze Age. It was during this period that international commerce expanded and wealthy rulers and aristocrats were buried with treasures of gold, silver and bronze.

Metalworking brought about great changes in the early societies where it was invented or introduced by wandering metal smiths who were always welcome wherever they went. Metals were agents of great change. They touched almost every aspect of human activity, upsetting the established order of Stone Age societies and leading to economic specialization. The discovery and use of metals led to the abandonment of total self-sufficiency by isolated Neolithic communities and stimulated trade of an increasingly complex nature.

Metals raised the standard of living, particularly in towns strategically located near gold or copper-bearing ores or along trade routes. Urban dwellers depended upon farmers for food. They offered raw materials, goods or services in return for grain, wine or wool. Metals made life more comfortable. Sophisticated city residents developed an appetite for a variety of imported luxuries which included metals. Prospectors began to search farther away for gold, copper, silver and tin. Traders plied the land and water routes connecting sites of metal-producing ores with those that needed raw ores, smelted ores or finished articles. By 4000 B.C. sizeable ships were exploring rivers and lakes. By 3500 B.C. Egyptian merchant ships were traversing the Mediterranean. These hardy sailors were the explorers and discoverers. Prospectors, traders and itinerant metal smiths followed, spreading techniques and ideas—the seeds of civilization—among scattered peoples in the Near East, Western Asia and Europe.

It was a relatively simple matter for prehistoric gold seekers to pan for gold dust and grains. The process involved placing some gold bearing sand or gravel in an open basket, hollowed out gourd or flattish dish and then swirling it around in running water with a deft motion of the wrist so that the gold settled to the bottom of the pan as the lighter material washed over the edge. In dry areas the auriferous sand was tossed into the air from large basket and winnowed the way the air carries off the chaff from grain. The light waste material blew away and the gold fell back in the basket.

Another method of gathering gold made use of sheep skins

and sluice boxes. They were fastened to the bottom of a flowing stream or held against the current so that bits of gold, which has an affinity for oil, were trapped in the lanolin-sticky fleece. This was particularly effective when heavy rains pouring down gold-filled mountain sides bore fresh bits of the precious metal into swollen streams. The Persian king Darius in the 6th century B.C. maintained a vast force of slave laborers to work when spring floods washed gold into rivers. Shifts of slaves tended the fleece-lined riverbeds catching light bits of floating gold which had escaped skins held taut upriver against the current. Camel caravans came and went carrying away the gleaming fleeces and supplying the gold-washers with fresh skins.

The Golden Fleece of Greek mythology was such a gold flecked skin. The practice of trapping gold this way spans history and was used by miners during the California gold rush who substituted rough wooden blankets for sheep skins.

According to Greek legend the first gold rush in history was initiated by Cadmus, a Phoenician prospector and king who went to Greece in the 14th century B.C. and settled in Thessaly where he found gold. He was said to have brought the alphabet to the Greeks and founded the city of Thebes. Legendary Jason, who sailed with his Argonauts to the eastern shore of the Black Sea in search of the Golden Fleece, was his descendant.

The legend tells how once in Thessaly there was a marvelous flying ram whose fleece of pure gold gave him magical powers. He had been a gift from the god Hermes to Nephele, who had sons and a daughter by the king. Because she feared her husband's concubine would harm her children, she begged the ram to carry them to safety in the kingdom of Colchis. The daughter grew dizzy and fell into the sea but the boy arrived and sacrificed the ram to Zeus. King Aetes of Colchis treasured the fleece above all things. He put it in a sacred grove guarded by a most terrible dragon because an oracle had warned that his life depended or not losing it.

Jason set out with a hardy band on the ship built by Argos to retrieve the Golden Fleece. With the help of Aetes' daughter, the sorceress Medea, Jason captured it. He took the fleece and the princess back to Greece. That is the story which probably has its origins in dimly recalled early gold-seeking expeditions to the Caucasus region of the Black Sea where gold was available to prehistoric traders.

The goldsmith was the first of the metalworkers and throughout history the aristocrat of smiths. He practiced a craft that was not only mysterious, but sacred, because he worked with the divine metal. Even now in certain West African tribes with a long tradition of fine gold work, only a chief may be a goldsmith. And in parts of Borneo and Sumatra where gold has been crafted for thousands of years, goldsmiths, until recently, were regarded with such esteem that during wartime they were neutrals who could pass unharmed through hostile territory. The ancient Jews believed that the art of working gold and other metals had been brought to earth by fallen angels who formed unions with Hebrew women to whom they taught their secrets.

The first applications of gold were of a magical nature, but those with power soon wanted gold ornaments for themselves to attest to their rank. It seems to be a part of human nature to decorate oneself and the connection between gold and personal adornment was fashioned early. Goldsmiths were delighted at how responsive the yellow metal was to simple techniques. The modern world is delighted to discover how early goldsmiths mastered more difficult procedures.

The frontiers of history are receding in the face of constant archeological discoveries that cast a new light on the remote past. Until about 70 years ago archaeologists thought that the people who lived in Anatolia, that part of Turkey which is in Asia, during the early Bronze Age were members of a rather primitive culture based on very simple agriculture. However, in 1935 archeologists made a startling discovery which turned former estimates of archaic Anatolian development topsy-turvy.

In the course of excavating a tell, or mound, at Alaca Huyuk, about 100 miles east of Ankara, a treasure of splendid gold works emerged. They came from a royal necropolis of 13 tombs dating ca. 2500 B.C. and stunned everyone with their variety, richness and profusion. They indicated that aristocrats had enjoyed a life of wealth and luxury based on trade during the time when Ur and Troy II, whose treasures were already known, had flourished.

The site lies at the junction of three of the world's earliest trade routes to Mesopotamia, to the Black Sea and to the Aegean—and was continuously occupied from ca. 4000 B.C. The first settlement was a simple community of farmers and a few traders. It grew into a larger town which supported a class of wealthy, accomplished people. The small cemetery where the treasure was found was the resting place for several generations whose culture was established throughout the Anatolian Plateau. The tombs, ranging from a date of ca. 2600 to ca. 2300 B.C. were located in the midst of city buildings. The bodies, buried individually or in male-female pairs, were placed on their sides in a flexed position in one corner of the shallow rectangular shafts that were lined with stones. Most faced south and were surrounded by religious and personal objects. The large amount of gold and silver in the burials of single women indicates their unusual high regard. They were furnished with rings, pins, bracelets, diadems, anklets and buckles of decorated gold and silver. Everyone had been given ample household and toilet articles including gold and silver goblets, chalices, jugs and bowls. A strangely beautiful pin of meteoric iron with a gold head was found in one burial.

The Alaca Huyuk artifacts were not crude pieces, but the product of sophisticated smiths who made elaborate use of sheet gold, gold wire for delicate filigree, repoussé and inlays of metal such as electrum, semi-precious stones and jade. The pottery found in the tombs, which was made in the town, is rough compared to the refinement of the metalwork. Most of the gold appears to have been worked in northern Anatolia and reached Alaca Huyuk in trade. The style of most of the gold, silver and bronze objects is almost purely Anatolian and shows the high level of metallurgical technique developed by the Anatolians. There is little influence from the Aegean or Mesopotamia where goldsmiths were working, although several gold pins were almost identical to pins found at Troy II. A few other sites in Anatolia have yielded grave goods of this period in precious metal and bronze, but none have been as splendid as those from Alaca Huyuk.

Across the Black Sea, rich burials of Copper and Bronze Age magnates have been excavated. In the fertile black soil of the Kuban Valley, only 350 miles from Alaca Huyuk, Caucasian aristocrats were buried in a royal cemetery at Maikop. Under large mounds archeologists found a wooden mortuary house divided into three chambers. A skeleton, sprinkled with red ochre, lay in the central room under a canopy with gold and silver supports. He was surrounded by a lavish display of gold ornaments, gold vessels and silver vases engraved with animal scenes.

Metallurgy had a precocious beginning in the Caucasus where gold, silver and copper abounded in its mountains. Maikop was a Copper Age cemetery of the early Kuban culture, from the middle to late 3rd millennium B.C. In the succeeding Bronze Age the practice of kurgan burials, so called from the Russian word for the mound covering the grave, spread to the south Russian steppes and the Ukraine. At Trialeti, between the Black Sea and the Caspian, such pits have been excavated containing bodies buried singly in four-wheeled carts and surrounded with a wealth of gold grave objects of superb workmanship. Some time after 2500 B.C. kurgan culture, which included the goldworking techniques and styles of the Maikop people and the domestication of the horse, was carried into Eastern Europe and eventually influenced most of Northern and Central Europe.

It is easy to overlook the achievements of Bronze Age Europe. The glories of Mesopotamia and Egypt and the sophisticated courts of the Minoans, Trojans and Mycenaeans tend to overshadow the cultures of the North. None of mankind's brilliant civilizations flowered there. It wasn't until Roman expansion in the centuries around Christ's birth that Europe, north of the Alps, entered the mainstream of urban civilization. However, there were scattered pockets of gold-producing areas where goldsmiths made articles reflecting their particular culture or copied the rare object of manufactured gold which somehow made its way from Asia Minor or even farther away.

Immigrants from Western Asia straggled into southeastern Europe early in the 6th millennium B.C. They introduced agriculture to scattered groups of hunter-gatherers and farming gradually spread north and west, reaching Scandinavia and the British Isles some time between about 4300 B.C. and 3500 B.C. These rudiments of cultivation were supplemented by three successive waves of information. The first was the early flow of technology from Greece and Anatolia to Eastern Europe and from there up the Danube River and its tributaries into the heart of Europe. Later, a second flow reached into Atlantic Europe from the Mediterranean, and, even later, migratory people streamed across the Northern European plains from the Black Sea to the Baltic bringing their technology with them.

Scholars once thought that metalworking developed in the Near East and was introduced to ignorant Europeans. New evidence, however, indicates that it was in part invented independently in eastern Europe where gold and copper enriched the Balkans. In the 5th millennium B.C., before farming had spread to the fringes of the continent, goldsmiths in the Balkans were working gold from Transylvania, and coppersmiths were turning out tools superior to contemporary ones made in Western Asia.

In the 4th millennium B.C. people of the Tripolye culture in the western Ukraine and eastern Rumania made cult objects and jewelry of gold exchanged in barter. They lived in towns that were a far cry from the cosmopolitan centers of Mesopotamia or the splendid courts of Egypt, but they were not as crude as historians once thought. Farmers cultivated grain and tended cattle. They lived in villages as large as a hundred well-made long houses, which were built on terraced lands near streams.

They made surprisingly fine pottery, painted in many colors and boldly decorated with geometric and curvilinear patterns.

Some of their gold has been unearthed at Cucuteni, the site of a village in Rumania.

Europe's early goldsmiths made do with the most primitive techniques. While goldsmiths to the south were crafting dazzling masterpieces, their northern counterparts made gold rope into coiled beads and rings. They hammered out thick gold foil and shaped it into beads or plaques which they embossed with rough designs, often no more than a pattern of raised dots. Small plaques in the form of goddesses from the 4th millennium were found at Russe on the southern bank of the Danube, and little perforated gold disks were excavated in Tibava in eastern Slovakia. They are among the very few surviving examples of archaic European gold. Other finds included beads and simple rings for fingers, ears and locks of hair. They have been unearthed from the sites of prehistoric communities which traded amber, salt, flint and furs along the river valleys of Europe for thousands of years. That trade accounts for rare bits of gold found where gold wasn't native.

It wasn't until the Bronze Age was well underway in Europe that parts of the continent became wealthy enough through trade to support goldwork of elegance. Goldsmithing got a real boost during the 3rd millennium when successive waves of various migratory peoples, collectively known as the Bell Beaker Folk, rolled over the continent. Named for their distinctive pottery vessels, the Bell Beaker Folk were important because they spread gold, bronze and copper-working techniques over much of Europe, which was still essentially in the Stone Age. Their origins are hazy. Some may have come from the Eurasian steppes, some from the Iberian Peninsula where there was already a well established goldworking tradition. Some scholars think one group came across from North Africa. In any case they were skilled metalworkers and clever traders.

For a while they were mobile; roving from one Neolithic farming community to another. They made weapons and tools for the more prosperous farmers and gold jewelry for the elite in a society that was rapidly becoming more sophisticated. Eventually they settled down with the local citizenry from Poland to Great Britain.

The hybrid European culture resulting from this mingling adopted such Bell Beaker practices as burying the dead in single graves under barrows and furnishing them with gold, amber, onyx and other luxury items gained in trade.

Thousands of years after real cities had grown up in the Near East, communities in Europe began to develop into more complex units. Society became ever more stratified. Very few Europeans could afford tools and weapons of cast bronze or even of copper, and an even smaller number could afford to commission gold ornaments from wandering goldsmiths. Life for the ordinary man remained much the same as he continued to use implements of stone, bone and wood. Metallurgy had an interesting effect on the styles adopted by European Bronze Age stone workers, who often made their stone axes imitate metal ones. Some were perfect replicas, so finely made they even had an imitation seam line where the halves of a metal axe would be joined in the casting process.

In towns along trading routes the number and types of specialists multiplied. Life became more interesting as exotic currents bearing new products, new faces and fresh ideas flowed through areas that had formerly received only occasional bartered goods from distant places. Regional chieftains became tremendously wealthy, heaping up profits from trade they controlled as prehistoric Europe was drawn into the growing network of trading routes that reached into the metal-rich zones of the continent and Britain. Long distance organization developed to profit from deposits of metals that found an insatiable market to the south.

Not long after 2000 B.C. Levantine merchants and smiths were in contact with the gold and copper-rich areas of south-central Europe. They traded for Cornish tin and Irish gold on the south coast of England where the rich and brilliant Wessex culture blossomed. During the 2nd millennium, as Egyptian, Minoan, Phoenician and Mycenaean power waxed and waned, a web of trade routes ran from Africa to the fringes of northern Europe.

During this time a group called the Uneticians, after a Czech village where a treasure of metalwork was found, made important contributions to European goldwork. They emerged from a center in the auriferous Carpathian Mountains after the Beaker people had settled down. They were farmers with an advanced knowledge of metals who spread their influence throughout the fertile valleys of Bohemia, Moravia, Silesia, Saxony, Bavaria and the Rhineland. Making use of procedures originally learned from wandering smiths, they built a vital bronze industry. They used both local and imported gold to supply the nearby aristocracy with jewelry. Gold from the Carpathian and Bohemian Mountains was supplemented with Irish gold.

They were great borrowers of stylistic ideas. Goldsmiths added to their repertoires elements of design, some of them already centuries old, from such goldworking centers as Troy, Syria and Sumer. Unetician bronze smiths made so many of a particular type of neck ring, which imitated a Syrian model, that they served as a kind of currency throughout a vast area. By the middle of the 2nd millennium the Uneticians were the dominant people of Europe. Their chieftains and advisors controlled commerce and practiced an early form of banking in fortified hilltop towns. Each town maintained a treasury where the community's stock of metalwork was stored, and an inventory was maintained recording all transactions.

At first the people of the Unetician culture and related groups buried their dead in cemeteries of single underground graves. During the Middle Bronze Age they adopted tumulus burial. The deceased was laid out on the ground, surrounded by personal possessions thought to be comforting or useful in the afterlife. A box-shaped structure of stone slabs was built around the body, and piles of dirt and stone were mounded over it. Sometimes the wife and children of a powerful regional chieftain were sacrificed to keep him company. These tumulus burials in Central and East Europe gave their name to the culture of warlike peoples, descendants of the Uneticians, who flourished from ca. 1500 B.C. to ca. 1200 B.C. Traces of their fine metalwork have been found in tombs and in secret hoards probably buried in wartime.

Archaeologists have excavated a number of large tumuli in the area between the Saal and Elbe Rivers in Central Europe, a crossroads of the salt trade in remote times. Later the area prospered from its ideal situation at the junction of several important trading routes. Local magnates grew wealthy trading in raw ores and through control of a bronze industry which turned out superb weapons. The most outstanding of the amazingly rich burials that have been found is a 16th-century B.C. tumulus near Leubingen in East Germany. First excavated in the late 19th century A.D., the mound was originally almost 10 yards high, 37 yards in diameter and 157 yards in circumference. Beneath a layer of later Slav burials was a layer of sterile soil about five yards thick. The ancient burial under it was covered with a conical heap of stones more than two yards high and 22 yards in diameter.

The stones covered a thatched-roof mortuary house of planking and clay. In it were skeletons of an old man and a girl about ten years old who lay across him at right angles. The chieftain had been decked out with fine bronze weapons and a beautiful stone adze of serpentine. The Bronze Age girl had been given gold earrings a massive bracelet of solid gold with a chased rib decoration, finger rings, dress pins and a spiral hair ornament, all of gold. Other burials contained gold disks and assorted gold jewelry including striking, large bracelets and rings for hair which were widely copied in bronze as far west as Ireland.

Aristocratic warriors were sometimes entombed with gold-decorated weapons. Treasure caches discovered quite accidentally in Hungary and Rumania have yielded rare solid gold swords and axes, cast in a mold similar to utilitarian bronze weapons. One site in Rumania contained a gold dagger, part of it missing, which weighed three pounds and was of high quality 21-carat gold.

These ritual weapons, obviously too soft for practical use, show the link between European Bronze Age goldwork and the earlier tradition of the Mediterranean world where such ceremonial pieces originated. Commercial contacts opened new horizons to the Europeans who were quick to adopt what pleased them, making modifications to suit their own cultural preferences. The pace of trade quickened with advances in mining European ores made by the Urnfield peoples, whose culture succeeded the tumulus period. This vital group of related cultures spread the practice of putting cremated bodies in pottery funerary urns as far away as Sicily, southern France and northern Spain. They were excellent metallurgists and made significant improvements in prospecting for metals and the deep mining of copper ores. In the Austrian Alps, for example, Urnfield miners who burrowed 400 feet into the mountains, opened more than 32 copper mines in one square mile area.

The great explorers and traders of the Mediterranean, the Mycenaeans, made a great impression on Bronze Age Europe, especially in goldsmithing. The mainland Greeks, famed for their war with Troy, had inherited mastery of the seas from the Egyptians, Minoans and early Phoenicians. They established trading posts around the Aegean and then ventured further west to the islands off Sicily. By the 2nd millennium B.C. they controlled the supplies of gold and other metals from Britain and continental Europe.

A few Mycenaean goldworks have been found in Europe along with many copies of Mycenaean styles. A magnificent gold cup found near Bonn in the Rhineland closely resembles one found in one of the shaft graves at Mycenae. Another lovely cup, also raised from a piece of sheet gold, which had been hammered thin from a single nugget, was found in a barrow burial at Rillaton in Cornwall. Its walls are fluted which lend a striking appearance and also strengthen them. To make it the smith probably filled the interior with sticky, black pitch as he carefully modeled the fluting and decorated the bottom. The handles of both cups are patterned strips of gold ingeniously fastened to the sides by gold rivets cushioned against gold washers. They are astonishingly beautiful pieces made by a master who must have trained in a sophisticated goldworking center. However, it is uncertain whether the cups were made by Mycenaean smiths in Greece and imported, or by European smiths trained by Mediterranean craftsmen.

Southern England became a major entrepot for international commerce as the demand for metals and other northern goods multiplied. Mediterranean seafarers with increasing frequency guided their ships out of the Mediterranean into the treacherous Atlantic and up to England's south coast. The fertile chalk downs of Wessex gave their name to the Early Bronze Age culture which thrived on the profits of trade. Raw gold, copper and tin, Baltic amber beads, and other northern goods were collected in the Wessex area and exchanged for bronze implements, faience beads, manufactured goldwares and other luxuries from the civilizations of Mycenae and Crete.

No Wessex settlements have been discovered, but archaeologists have found ample evidence of the great wealth and far flung connections of the English traders in over a hundred rich graves. There is one other outstanding indication of the might of the Wessex chieftains. It was they who were responsible for the last phases of Stonehenge—the greatest of megalithic monuments—so large that the entire city of Troy II excavated by Schliemann, could easily be set inside the ditch running around its perimeter.

Around Stonehenge the Wessex people managed to add sandstone monoliths—some weighing over 50 tons and brought from 24 miles to the north—is a complex of ritual sites and barrow cemeteries. Fortunately for posterity, those early English princes of commerce believed in taking their gold with them to the grave, unlike Irish chieftains and Bronze Age leaders in the Carpathians who flaunted their golden wealth only as long as they lived.

The Wessex burials contain bronze halberds and pendants from East Europe, amber from the Baltic shores, pins from the Unetice culture of Central Europe, faience beads from the Mediterranean and gold. There was gold jewelry, gold-hilted weapons of Aegean bronze, gold beads, amber beads bordered in gold, scepters of gold and inlaid with onyx which seem to have been inspired by Mycenaean ceremonial maces, Irish crescent-shaped gold necklaces, belt buckles, twisted gold earrings from Cyprus and shining hair ornaments from Ireland and Egypt.

The largest, most impressive surviving piece of British Bronze Age gold is the Mold Cape, a fragmented wrapping of ornamented sheet gold found draped around a skeleton in North Wales in 1833. The cape was made from a single piece of thick gold probably beaten to shape over a form. Strips of bronze were fastened to the back to strengthen it, before it was probably sewn on a cloth or leather backing, long since rotted away. The magnificence of the cape comes not so much from its size but the lavish surface decoration. The cape is covered with an elaborate pattern suggesting rows of variously shaped beads separated by parallel ribs. Similar profuse repoussé sections appear on a small number of gold articles discovered in central and northern Europe including two peculiar conical "hats" and a post-like object of gold leaf from Germany.

Britain has a long tradition of goldwork. The Beaker colonists introduced metalworking to the islands between 2500 B.C. and 1800 B.C. Ireland was especially fortunate because the streams of County Wicklow were so filled with gold that they weren't exhausted until recently after some 4,300 years of washing. Irish goldwork became famous. Early smiths of the Food Vessel people in Ireland and Scotland made magnificent crescent shaped chest ornaments called lunulae, which were traded to the Wessex people and found their way across to Northern Europe where they influenced continental goldsmithing.

Northern European fashions were in turn reflected in Irish gold. About 800 B.C. smiths in Ireland began to make broad neck ornaments called gorgets, which were based on continental pieces. They were of sheet gold, decorated with raised ribs and repoussé patterns. At either end a round terminal piece, slightly depressed like a saucer and beautifully embossed with delicate tracery, was attached. Another adaptation of style was a type of continental garment fastener. They have been found in hoards, never in burials, so how they were worn is open to speculation. One large fastener found at Clones consists of an arched bar with a big cup-shaped terminal at each end. The cups probably fitted through two loops at the point where a cloak joined in front and held it together.

With the collapse of Mycenaean civilization at the close of the 13th century B.C., the Mediterranean world no longer absorbed great amounts of gold from Britain or the continent. Nor did Mediterranean wares reach the north as they had earlier. Since gold no longer drained south, the goldsmiths in auriferous areas were able to manufacture more articles in keeping with local taste. Gold drinking cups, for example, became fashionable in much of Europe during the Late Bronze Age. Treasure hoards from such places as Romania and Hungary contain an unusual number of barbarian gold cups and gold diadems and jewelry of purely local inspiration, showing that when Mycenae fell and contacts faded, people returned to a more isolated existence, at least until the dawn of the prosperous Iron Age.

One area that embarked rather late on the Bronze Age was Scandinavia where some of the most remarkable early gold has been found in graves, treasure hoards and bogs. The Stone Age lingered on in Scandinavia and the plains of northern Europe, which were not endowed with metal deposits. While the rest of Europe was already enjoying Bronze Age technology, Neolithic northerners managed to acquire gold through trade in furs and especially in amber, the precious sea-gold picked up along the shores of the Baltic and North Seas. Amber was treasured from earliest times for its glowing beauty, fragrance and alleged magical powers. Bits of the fossilized resin of very ancient conifers have been found in Stone Age graves as far south as Spain, France and Moravia.

After the middle of the 3rd millennium B.C. amber, associated in the north with sun worship alike gold elsewhere, flowed south along the amber roads spreading throughout Europe and Great Britain, reaching as far as Mesopotamia. Goldsmiths everywhere used it in conjunction with gold to fashion beads, buttons, brooches, rings and inlays for weapons and ceremonial objects. The amber trade brought southern Scandinavia into its own Bronze Age about 1500 B.C., while hunting and fishing Neolithic cultures continued further north. The Danes benefited most since all traffic in amber and furs passed through their hands. Ingots of copper, tin and gold were imported along with finished articles. Gold was suddenly more plentiful, and goldsmiths whose grandfathers had worked in flint developed an independent tradition in gold that was the equal in quality to that of other European areas.

The aristocracy of Denmark and southern Sweden sported gold ornaments, made golden votive offerings to the gods and buried their dead with warm glowing gold in massive burial mounds.

The first gold to come so far north was already worked. The first efforts of Scandinavian smiths were clumsy and crude imitations of imported articles. Within a couple of centuries, however, aided by foreign goldsmiths probably from Central Europe, they became highly skilled at casting gold. Still, they never perfected the art of hammered goldwork, in which Central European smiths excelled, although they loved the golden bowls raised from thin sheet, caressed into a variety of shapes and ornamented with repoussé patterns of delicate ribbing, beading and concentric circles. Countless of these pieces were imported and many have been recovered from peat bogs where they may have been thrown as votive offerings. On some of them Danish goldsmiths added a peculiarly Scandinavian touch to satisfy local taste. They took an imported bowl, for example, and added strange horned horse heads with long curved necks which served as handles.

The most exciting finds of Scandinavian gold have come from the bogs. In addition to the imported cups and bowls, offerings included a whole fleet of miniature golden ships, ribbed gold neck rings, massive armbands and the characteristic goldwork of twisted gold bars and coiled gold strips made into bracelets, earrings and pins. The spiral was a popular motif and may have been associated with eternity because of its endless coiling.

It is not surprising that the people in the northern latitudes were sun worshippers. One of the most impressive treasures of Bronze Age gold is the famous Trundholm Sun Carriage found in a peat bog in Denmark. It is a miniature bronze carriage bearing a sun disk and pulled by a horse, and was locally made although influenced by southern tradition. The wheels are of the four-spoke type used in Egypt and Mycenae for ritual model carriages long before 1300 B.C. when such objects made their first appearance north of the Alps.

The splendid Scandinavian Bronze Age, late to flower but outstanding in the achievements of its gold and bronze work, lasted until about 500 B.C. Then the prosperous age ended as the climate became much cooler and damper. During the same period Celtic dominance was established in Central Europe, severing Scandinavian trading links with the south. In addition, the bronze-based economy of Scandinavia was severely disrupted by the introduction of iron weapons and tools from the continent, which had already embarked on the Iron Age.

At the turn of the 1st millennium, the Celts, whose culture evolved in Central Europe from that of the vigorous Urnfield peoples, initiated Europe into the use of iron. Their carbon-hardened iron weapons with keen, durable edges quickly made the Celts masters of the continent. Bronze swords—brittle if they contained too much tin, soft if they had too little—were no match for the new weapons. Within 500 years the Celts held sway over a vast area from the Black Sea to Ireland's Atlantic coast.

Celtic art, so beautifully expressed in goldwork, was the first conscious style born north of the Alps. The distinctive Celtic style was applied to works in wood, stone, glass and pottery, but reached its peak in metalwork. Personal ornaments, vessels and religious articles in gold, silver and bronze were adorned with highly stylized forms quite unlike the naturalistic motifs applied to goldwork of the same period in the Mediterranean area. Celtic taste emphasized abstract patterns and stylized human and animal figures.

The earliest examples found come from princely tombs on the middle Rhine, the zone that was the center of late Hallstatt and La Téne activity. The Austrian site of Hallstatt has given its name to the principal early Iron Age culture in Central and Western Europe which lasted from about 700 B.C. to around 500 B.C. when it was succeeded by the La Téne Phase, which takes its name from a site on Lake Neuchâtel in Switzerland. This second period of the

European Iron Age ended only when Celtic culture was ended by the Roman conquest. The Romans never invaded Ireland, which remained a sanctuary for Celtic culture and art into the Early Christian period.

The Celtic goldsmiths who practiced their craft from Hungary to Ireland produced some of history's most stunning jewelry in the form of torcs, arm bands, rings, pins and helmets. They were first class technicians and confidently employed a wide range of techniques including casting, repoussé, enameling and inlaying with coral and other materials. The splendid Celtic tradition incorporated influences from four sources: a local Hallstatt style, an animal style from the Eurasian nomads, elements from the Etruscans and Greeks and an oriental element.

The Celtic peoples were politically divided but shared a common culture with those living closest to the Mediterranean. They developed a more sophisticated society because of contact with civilizations of the Etruscans in Italy and the Greeks, whose influence was widespread in the centuries before the Roman rise to power. Iron Age society was highly stratified with chieftains and a class of aristocrats amassing wealth and stimulating the production of precious golden possessions. Craftsmen filled their orders; peasants produced food for everyone and slaves performed a variety of tasks.

The wealthy chieftains organized the export of metal ores, salt, amber, furs and other products and received cherished luxury items from the Greeks and Etruscans. Aristocrats were often buried with wagons or chariots and a profusion of grave goods, both domestic and imported, including jewelry, richly decorated weapons and ceramic and metal articles associated with wine drinking, a Mediterranean practice which had become very fashionable among the barbarians.

Wagon burials tell a lot about Iron Age goldwork, both its appearance and the relative amounts available at different periods. Chieftains' graves of the 7th- to mid-6th centuries B.C., for example, contain little gold, which indicates its scarcity. However, after that time Europe, north of the Alps, began to trade with the early Classical world. After the Greeks established a colony at Massilia (Marseilles) in southern France, much more gold appears in burials. This happened as trade that had long followed the prehistoric amber roads through Central Europe to the head of the Adriatic now shifted westward to routes running from Massilia along the valleys of the Rhone and Saone and from there to the Danube and Rhine.

The most outstanding Hallstatt burial was excavated at Vix in northern France. The mortuary house of a Celtic princess whose family seems to have ruled an Iron Age hill fort nearby was filled with treasures. When the princess died about 500 years before Christ was born, she was about 30 years old. She had been placed on a chariot draped with blue and crimson cloth embroidered with small bronze plaques. On her head was a solid gold diadem and she was decked out with a variety of gold ornaments, some of local origin and others imported from the lands of the Scythians, from Greek workshops and Etruscan territory. The most splendid item in the Vix burial wasn't gold at all but bronze—a massive bronze krater, or mixing bowl for wine. It was made by Greek artisans and is over five feet high and weighs 457 pounds, which makes it the largest metal vessel to have survived antiquity.

The story of gold in prehistoric Europe is not as dazzling as that of the sacred sun-metal in the brilliant centers where high civilizations first flowered. But it is nevertheless, a very interesting story and a much older one than scholars once thought. Gold was rare and life, for the most part, was hard in Europe. The courts of warrior chieftains who profited from expanding international trade were crude in comparison with those polished urban centers which shone along the Nile and between the Tigris and Euphrates or on the shores and islands of the Mediterranean. Perhaps, however, in the northern climes where skies were often cloudy and the earth was shrouded frequently in mist or rain, what gold there was gleamed all the brighter with its eternal magic promise of light, life and power.

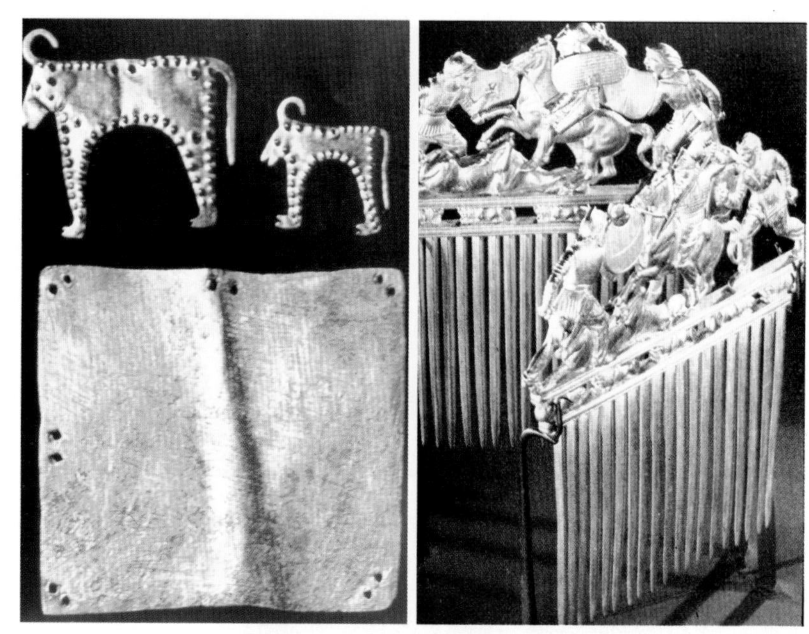

(Above, left) Thracian gold jewelry—oldest European gold—4th century B.C.
(Above, right) Scythian gold comb of Greek manufacture from 5th century B.C.

Sumerian jewelry from Ur—necklaces of gold, lapis lazuli, and carnelian—ca. 2500 B.C.

Inca, Moche culture gold ear ornaments.

Scythian gold plaque, 7th century B.C. from collection of Peter the Great depicting a monster and tiger in combat.

Map showing changes in Assyrian Empire from 824 B.C.–671 B.C.

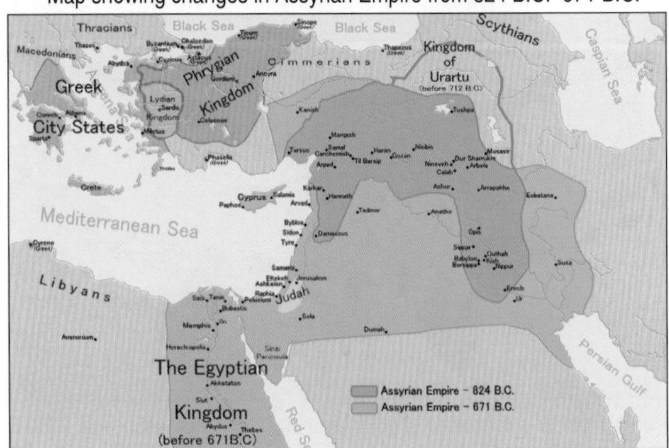

(Above) Minoan gold pendant of nature-god ca. 1650 B.C.
(Below) Tartessian gold jewelry ca. 6th century B.C., discovered in a grave near Seville Spain.

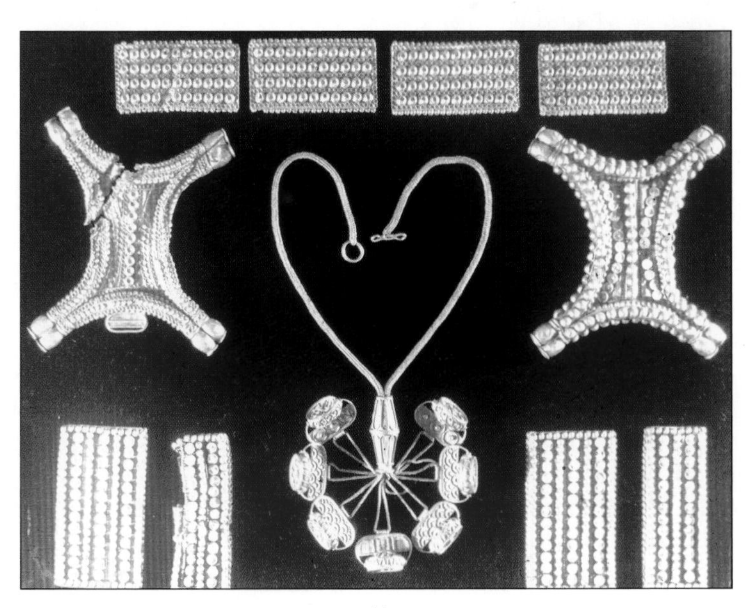

Tomb of King Darius II Nothus (423–404 B.C.)—near Persepolis.

CHAPTER 3

MOUNDS, MESOPOTAMIA AND MONEY

In the Middle Ages a few Jews and Arabs, the most daring of Medieval travelers, journeyed to the bleak alluvial plain of what is now Iraq. They brought back curious tales of strange mounds of earth which rose here and there breaking the monotony of the parched land between the Tigris and Euphrates Rivers. Tribesmen in the sparsely settled area told them that the commanding mounds were not natural hills at all. Local legend told of magnificent palaces and temples, filled with golden splendor, which were hidden beneath the earthworks. As they looked at the rude clusters of huts that made up the 12th-century farming settlements, such stories seemed preposterous. How could the ancestors of barbarian villagers have been sophisticated residents of a golden metropolis?

Centuries later a few Europeans began to venture into the area. A 16th-century traveler, Leonard Rauwolff, returned confident that there was indeed some truth to the ancient legends of golden cities. He felt sure that two rubble-strewn hills on the east bank of the Tigris marked the site of Nineveh, the Assyrian capital so frequently mentioned in the Bible. "Spoil is taken, spoil of silver, spoil of gold; there is no end to the store, treasure beyond the costliest that man can desire," wrote Nahum in the Old Testament oracle about Nineveh, one of antiquity's most splendid cities. But neither Rauwolff nor anyone else in the succeeding centuries really explored any of the mounds and the world remained unaware of the amazing secrets cloaked in sand and rubble.

Apart from Biblical references and vague remarks in Greek and Roman histories, no trace of Mesopotamia's gold and glory remained, save for the rubble mounds. Ur of the Chaldees, Babylonia and Assyria were familiar names because of the Bible. But most of the Biblical descriptions were discounted by scholars of the 18th century. And no one even suspected there had been a nation called Sumer; the world's first civilized state, where, in cities of the late 4th and early 3rd millennia B.C., mankind scored a series of dazzling achievements, including invention of the wheel and of writing. The gleaming palaces, the glazed brick towers, the temples filled with gleaming treasure were subjected to the ravages of war and time. While the great stone monuments of Egypt endured, the colorfully glazed mud-brick buildings of Mesopotamia's almost 3,000 years of history crumbled. Rain, flood waters and blasts of gritty desert wind worked on the wondrous monuments until even the highest of the gleaming ziggurats, the famed Tower of Babel, toppled. Villagers carried away bricks and rubble for their own buildings in the stone-poor land, and little by little the nations and people who had "praised the gods of gold" were forgotten.

It wasn't until the 19th century that the ancient mounds were excavated. Then, layer by layer, foot by foot, archaeologists discovered the glorious accomplishments of Mesopotamia. While treasure hunters were flocking to the grandiose monuments of Egypt, a few dedicated British and French diplomats began to unravel the mysteries of the rubble-strewn hills. They were not professional archaeologists and had only the simplest methods for recording and preserving their finds, but they made a number of significant contributions to the fledgling science of archaeology.

The greatest of these men who combined a civil-service career with a love of the past, was Austen Henry Layard. He was an exuberant and determined Englishman who didn't let little things like being robbed time and again or being briefly held as a slave by Bedouins, diminish his digging enthusiasm. His mid-19th century excavations at Nimrud, the Assyrian military capital of Kalhu, uncovered the remains of three royal palaces in the first month.

The greatest of these was a six-acre palace, once filled with treasure, which had been built by the cruel tyrant Ashurnasirpal in the 9th century B.C. Layard and the world were awed by the maze of royal apartments, audience halls and ceremonial chambers built by a king whose very name had struck terror in the hearts of people in many countries. His armies ranged far afield in search of spoils. He held sway over a vast area, receiving tribute in gold and other precious materials. His cities and palaces were filled with gleaming metals, precious stones, furs, ivories; incense and rare perfumes scented the air. His women and courtiers dressed in gold-embroidered silks and bedecked themselves with splendid jewelry. But Layard found none of these treasures, vanished with the passing of Assyrian might. Nor did he find much gold at Nineveh, the Assyrian capital on the Tigris, where he peeled off layers of history like onion skins. Instead, he found a far more valuable treasure—the libraries of Sennacherib and Ashurbanipal, two of the mightiest Assyrian monarchs.

At almost every Mesopotamian site clay tablets were found in abundance; sometimes thousands of them with "funny" markings earlier dismissed as meaningless. But when scholars learned to decipher what resembled "bird tracks on wet sand." Mesopotamia suddenly came to life. The Sumerians invented writing about 5,500 years ago, and they and their successors in the land between the rivers evidently enjoyed writing, judging from the quantity of tablets found and their subject matter. Scribes set down every possible detail of Mesopotamian life from the exchange of gold with foreign leaders to the behavior of delinquent schoolboys.

Scholars deciphered cuneiform script tablets dictated by Ashurnasirpal to his scribes. He boasted at great length of his exploits and the atrocities he inflicted on the vanquished. "I flayed all the chief men," he said, "I cut off the limbs of the officers... Many captives I burned with fire," and so on. The conqueror who destroyed human life so callously was enormously proud of what he created. Speaking of his palace at Nimrud, Ashurnasirpal dictated, "I built a palace for my royal dwelling and for my lordly pleasure. Beasts of the mountains and of the seas...I fashioned and set them up at the gates...I made it glorious...silver, gold, tin, bronze and iron, the spoil of my hand from the lands which I had brought under my sway, in great quantities I took and placed herein."

He also described the colossal housewarming he gave. He invited some 70,000 guests from many countries to the new palace. In ten days of celebrating they polished off 2,200 roast oxen, 16,000 sheep, countless other meats, birds, eggs and fish. They downed rivers of beer and wine as they marveled at the sumptuous palace and the opulence of its furnishings and appointments aglow with gold, silver, ivory and precious stones.

Throughout Mesopotamian history despots coupled wholesale destruction of their enemies with ambitious domestic projects. They developed complex irrigation systems, built countless temples and palaces and lavished great care and a fantastic amount of gold on such urban renewal projects as Nebuchadrezzar's Babylon. Archaeologists have found many of the fabulous Mesopotamian sites but one of the most celebrated, Agade, the gold-filled capital of Sargon the Great, has so far eluded them.

Somewhere along the Euphrates in south-central Mesopotamia Agade lies beneath the gray desert plain. All that remains is a tantalizing reference on a clay tablet which reads: "The dwellings of Agade were filled with gold, its bright shining houses were filled with silver...Its walls reached skyward like a mountain." Sargon, founder of the mighty Akkadian kingdom in the mid-3rd millennium B.C. rose from cup-bearer to the king of Sumerian Kish, a conqueror who extended his dominion as far as the mountains near the Mediterranean coasts of Syria and Anatolia. He was the first leader to unite Sumeria with northern Mesopotamia. Four times his armies raided Palestine. Sargon built on a grand scale and was in constant need of fresh supplies of metals, stone, wood and the luxurious materials used to enhance his constructions.

The force which drove Sargon beyond his river valley was the need to control areas rich in the mineral and other resources Mesopotamia so sorely lacked. The same motive stimulated commercial activity and impelled the leaders of Mesopotamia to make treasure hunting-expeditions as well as military conquests throughout the 27 centuries of Mesopotamian history.

The lure of gold is so strong that even in a land where it is not found gleaming in river sands or underground, people desire it above almost all else and contrive to possess it. Mesopotamia was not endowed with native gold. Yet, from very early times its people valued the shining metal, acquired it through trading surplus agricultural bounty, and learned to work it in wondrous forms. The Sumerians, Mesopotamia's first civilized people, established a dozen or so city-states along the two rivers in the area between modern Baghdad and the Persian Gulf. In these wealthy cities the earliest merchant class evolved complex systems of international trade in which gold played an important role as it did in religion, art, economics and politics.

Sumerian cities evolved from farming settlements among the marshes. The land was barren of stone and mineral resources. The climate was harsh, with alternating searing heat and chilling winter winds. A most unlikely place for the flowering of civilized society. Yet the cooperation of men channeled the waters to irrigate the grudging land which produced an abundance of barley and other foodstuffs, and sustained large flocks of goats and sheep whose hides and wool, both raw and woven into excellent textiles or made into leather, was traded for gold and utilitarian metals.

The riches of flock and field were the wellspring of Sumerian culture which lingered thousands of years after Sumerian power was eclipsed by the might of Babylonia in the marshy southern half of Mesopotamia, and Assyria in the sandy north. The security of life enjoyed in a civilized society based on organized agriculture allowed men to address themselves to questions beyond mere subsistence. They experimented to produce the first recorded government by elected rulers, which by 3000 B.C. had given way to dynastic kingships; a numerical system based on 60, which we still use in telling time; the first astronomical calculations; the first compilation of laws; and great advances in goldsmithing.

Each of the Sumerian city states scattered on the dusty plain patched with green fields fertilized by river silt struggled for dominion over the others. Ur, Kish, Lagash and the others were linked only by language and culture. They cheerfully raided each other, carrying off prisoners and the gold lavished on temples and palaces. From time to time they all bowed under the yoke of a foreign master who stripped them equally.

In return for agricultural goods the Sumerians acquired gold, which the Mesopotamian's called "the strong, shining and enduring." It came to them from Arabia, the Caucasus, Turkey, Egypt and perhaps from as far away as the Indus Valley where there was abundant gold and an advanced civilization from about 2500 to 1500 B.C. By the beginning of the 11th millennium B.C. the reed huts and simple mud brick shrines of the first villagers had given way to a sophisticated society in which gold had begun to appear in the market place as a form of exchange. The life of the community centered around the temple complexes, which became increasingly elaborate as the numbers of gods and their priestly representatives multiplied. Each deity required daily ceremonial rituals, votive offerings, sacrifices and services. Nothing pleased them nor their priests and priestesses as much as gold.

Mesopotamian temples became the great centers of wealth and power. Their colorfully glazed ziggurats rose upward linking heaven and earth, dominating the bustling cities and fertile fields laced with canals. The kings were considered deities whose well being and comfort was crucial to the fortunes of the nation. Gold, the sun's own stuff, belonged to them and their fellow gods. Golden treasures were stored in the temples. It was in these great temples of Mesopotamia that the world's first banking, accounting and writing systems developed—all to protect and support the profits from tribute, taxes, trade, offerings and the revenue from temple lands. To protect the treasures piled up before the altars, treasure strong rooms—the first bank vaults—were built. They held accumulations of precious metal in ingot form and ecclesiastical treasures such as the ritual furnishings, ornaments and accessories of gold, golden textiles made into hangings, canopies and precious embroidered vestments required by the gods. Accountants kept track of the amassed wealth.

One goldsmith's plea for access to gold stored in the treasury is preserved on a cuneiform tablet. He had been ordered to make jewelry for a golden statue of the goddess Sarpanit but hadn't been given the gold he needed. "Everything has been deposited in the treasury of the temple of Ashur, and no one can open it except in the presence of the priest of X...Will the king be graciously pleased to send someone duly authorized to open the treasury so that I can finish the work and send it to the king?" The clay records fail to tell whether he completed the ornaments or not. Goldsmiths were highly esteemed for their skill. They were attached to the temples and palaces and furnished with living quarters, food and clothing so that "no time should be lost from their work."

In spite of strict security measures thefts sometimes occurred. A cuneiform tablet records one robbery in which the culprit was a temple craftsman. "The golden tablet which is missing from the temple of Ashur has been seen in the possession of the sculptor X... The king," dictated an official, "should take steps to have him sent for and questioned."

Gold stored in the sacred buildings was thought to enjoy special divine protection. Thus, when ownership of gold spread to the rich upper classes who provided Mesopotamia's generals, officials and priests, temple vaults began to serve as communal banks. Private gold was left on deposit for a small fee. Babylonian tablets record loans from the temple treasuries as early as 2000 B.C. Later, the Hebrews loaned gold from their temple vaults. They charged interest to cover the possibility of default on the loaned metal. In Greek and Roman times temple treasuries were of enormous importance, serving as official national depositories, lending money and arranging mortgages. Bank architecture, right into the 20th century, has reflected banking's temple origins. The Mesopotamians even had letters of credit. Early in the 2nd millennium B.C. priests from the temple of the Moon at Ur traveled with such letters written on clay tablets. These were exchanged for supplies and could be returned to the temple where the bearer would receive the amount of gold or silver specified in cuneiform characters.

Ur of the Chaldees is given in the Bible as the birthplace of the patriarch Abraham. When Layard and his contemporaries were digging into the Babylonian and Assyrian past, they never even suspected the existence of the far older Sumerians who founded Ur. In fact, it wasn't until barely a century ago that cuneiform clues hinting at a Sumerian culture were corroborated by excavation of the site of Lagash, one of the oldest centers of Sumerian civilization. Since a French consul in Basra discovered Lagash in 1877, hundreds of thousands of clay tablets and artifacts have been revealed by the archaeologists' diligence and have brought into clearer focus the first hazy glimpses of luxury and might of a magnificent civilization which spearheaded almost 3,000 years of Mesopotamian history.

Shortly after the end of World War I an English archaeologist, who had already worked in Egypt, began digging along the banks of the Euphrates, about a hundred miles inland from the head of the Persian Gulf. Sir Leonard Wooley had no idea when he began sifting through the layer upon layer of sand, sherds, ash and rubble at the site of the legendary Ur of the Chaldees mentioned in the Old Testament, that he was going to make one of archaeology's most spectacular finds. He excavated around the ruins of a once towering ziggurat and discovered an ancient cemetery.

That was in the fall of 1922. Instead of exploring it immediately, Wooley made the difficult and disciplined decision to postpone excavation until he had finished the area where he had previously begun work on. It was four years later that he used a hundred native diggers, to plumb the depths of the burying ground. It contained approximately 1,800 burials, on different levels and from different periods. Working slowly and with the greatest care, the distinguished archaeologist and his wife and crew excavated one tomb after another. In the third year, Wooley made a riveting discovery—the so called Royal Graves where princely Sumerians had been buried with gruesome ritual. These stone-lined, shaft graves were filled with a glittering array of golden treasures. The story they told revolutionized the knowledge of Mesopotamian pre-history, substantiated legends of ancient origin and revealed the high accomplishment of goldsmiths who practiced their craft 4,500 years ago.

The wealth of dazzling treasures, their elegance and refinement, coupled with the macabre proof of bizarre burial rites, which involved human sacrifice, have made the Royal Cemetery of Ur only slightly less celebrated than King Tutankhamen's gold-filled tomb. Actually, the Ur finds furnished more information about early Mesopotamian history than Tut's tomb did about Egypt of the Pharaohs. Wooley's dedicated spadework in the graves brought to light an amazing amount of gold. Far more, of course, was found with the young Egyptian king but Egypt was rolling in native gold and Sumer had none of its own. Yet, somehow, dozens of centuries ago the Sumerians were wealthy and organized enough to acquire gold in trade and through conquest.

Buried deeper than less rich graves to foil the tomb robbers who were active even in that remote era, the royal graves contained an array of grave goods that amazed the world, not only with their relative abundance, but their refinement and superb workmanship. They mark the technologically most important find in goldworking because they show that by ca. 2700 B.C. most of the technical processes used in working gold had already been mastered. In a "debate" inscribed on clay tablets written 20 centuries before Christ by a Sumerian poet, the personification of copper taunts precious silver, saying, "Silver, if there were no palace, you would have no place to be assigned to! Like a god you don't put your hand to useful work...Get into your dark shrines; lie down in your graves." How fortunate silver and sacred gold in such quantity did find its way into the graves of those who had lived in the Sumerian palaces. So much of what we know of prehistoric life comes from the "place of escape" as the poet called the grave. Men and women of high status, no matter how rich and powerful, turned to skeletons, but their dazzling artifacts emerged from thousands of years of darkness as fresh and lovely as when they were entombed, to cast new light on remote antiquity.

It seems incredible that Sumerian goldsmiths, who depended on imported raw material, could have crafted such perfect pieces. It is even more amazing that they became virtuoso craftsmen who could export their expertise to other nations, including Egypt, which had so much gold and such a long tradition of goldsmithing. They were among the greatest goldsmiths in all history, teaming technical sophistication with exceptional artistic sensibility. Archaeologists have found a number of goldsmith's shops which contained evidence that the Sumerians practiced applied chemistry in the refining of metals and in compounding unusual alloys.

Purity of gold found at Ur varies. Some analyzed pieces were of seven-carat gold or had 291 parts per 1,000. Others had 910.6 parts per 1,000, or were 22-carat gold. The Mesopotamians themselves tested imported gold for purity. A 14th century B.C. Babylonian king complains in two surviving letters to Pharaoh Amenhotep IV about the inferior quality of gold the Egyptian king, Mesopotamia's chief supplier, had sent him. He diplomatically implies that the Pharaoh himself must have been ignorant of the adulteration of the gold which weighed 20 minas when it went into the king's refining furnace and a mere five when it came out of the fire.

A breathtakingly beautiful gold helmet was found in the tomb of Mes-kalam-dug, a Sumerian warrior. It demonstrates the superb skill of the confident Ur goldsmiths in the difficult art of gold holloware. Found attached to the warrior's rotted skull, the helmet forms an elaborate wig with cheek pieces like flaps to protect the face. It was beaten up from a single nugget. The locks of hair are sculpted in repoussé. Each strand is incised to create an engraved pattern of waves and curls. There is a repoussé bun at the back of the head. A diadem encircles the brow, beaten in relief, as are the delicately modeled ears. No novice could have made such a piece. The master smith who produced it 4,500 years ago must have been the heir to many generations of experience.

Crafting gold holloware is extremely demanding because the gold, usually filled with soft bitumen or pitch, must be repeatedly annealed as it is thinned and shaped. The Ur smiths appreciated the inherent beauty and luster of gold. The surface of many of their works has been left undecorated to permit emphasis of the natural loveliness of the metal and the gracefully proportioned forms.

Among the most beautiful of the thousands of gold articles from the royal graves are those found with Queen Puabi, also known as Shub-ad. She was buried in the same pit as her husband who had died earlier. His sepulcher of stone had been robbed by the workmen preparing Puabi's tomb who had carefully resealed it to- cover their crime. In the large pit Wooley found the shocking remains of human sacrifices—64 court ladies, four harpists and various other retainers, all in ceremonial dress and wearing gold jewelry and metallic ribbons. Others of the 16 royal graves had as few as six or as many as 80 skeletons of attendants who appeared to have gone willingly to the hereafter with their royal masters who were considered gods. In none of the graves was there any sign of violence. Instead neat rows of women or composed figures of men lay where they belonged in life—grooms by their oxen, guards near their posts; all probably drugged by a narcotics and senseless when the pit was filled with earth.

The queen lay on a wooden bier in a limestone and brick chamber with two faithful ladies-in-waiting crouched at her feet. Her robe had long since disintegrated but she still wore a magnificent cloak of gold and silver beads worked with semi-precious stones and gold amulets in the form of fish and gazelles. On her skull was a stunningly beautiful head-dress, flattened by debris. Wooley lovingly restored it, puzzling out the hundreds of bits of paper thin sheet gold which made up elaborate blossoms and leaves and the beads of blue lapis, white calcite and red carnelian. Composed of three complex wreathes of gold ribbon, pendants, leaves and petals interspersed with colorful stones, it was evidently made to fit over a thick wig. A bouquet of gold flowers, swaying on supple stems rises from the top of the head-dress. Large double crescent-shaped earrings balanced the effect of so much head gold and Puabi wore other elaborate jewelry as well. A variety of rich gold offerings lay about her including a golden dagger with a 14-inch gold blade and a hilt of the bluest lapis lazuli set with gold studs. The blade was sheathed in a gold scabbard worked in a delicate openwork pattern reminiscent of woven grass. The queen had been provided with a second crown, which included among its elements thousands of minute beads of lapis lazuli and a variety of golden fruits and animal figures. Sumerian women used a variety of cosmetics to enhance their dark beauty, and Puabi's grave yielded a delightful collection of little gold pots containing make-up, sea shell cosmetic containers and seashell replicas in gold and silver for unguents and dyes. A delightful personal grooming set in pure gold which she must have used in her lifetime contains a tweezers, a head scratcher and an ear-scoop. Repoussé, hammered hollow-ware, sheet gold, granulation, filigree, simple open mold casting and challenging three-dimensional casting in closed molds using the lost-wax process—the Sumerian smiths excelled at them all. They seem also to have been the first to invent the process called cloisonné in which strips of gold were soldered onto gold backing to form tiny compartments. These little cells or cloisons were filled with semi-precious stones or iridescent shell cut to fit and cemented into place. Later, the cloisons were filled with colorful enamels or cut glass to make richly ornamented gold jewelry.

Ownership of gold was not restricted to royalty and priests during much of Mesopotamian history. Gold was used as a medium of exchange, and men and women adorned themselves with shining ornaments. When an Assyrian woman married, she was often presented with a gift of gold, albeit a small one, which remained her property even if divorced by her husband. Women and men too adorned themselves with shining ornaments and displayed wealth and status with gold personal effects. Among the goldsmiths' favorite motifs were rosettes that were associated with the sun. Rosettes, floral designs and the heads of ibex, cattle and lions decorated long pendant earrings, rings and heavy bracelets.

Gold from half the world away was drawn to Mesopotamia where it was crafted into jewelry and dazzling plate.

Trade, extensive and highly organized, was what attracted gold to the cities between the rivers. The powerful city state of Lagash, contemporary with Ur, was typical of the Sumerian cities whose early commercial activities attracted a continuing supply of raw materials and luxury goods from faraway sources. Lagash was a vibrant center ruled by independent kings in the 4th millennium B.C. Later under such Akkadian emperors as Sargon in the 3rd millennium, Lagash remained a great cultural center engaged in extensive foreign trade. Cuneiform records show that cedars were brought from the slopes of the Lebanon mountains and gold and copper from central and southern Arabia and the Sinai.

Various Mesopotamian kingdoms extended their dominion as far as Persia, Syria and Anatolia to control supplies of precious materials. Mesopotamian merchants, spurred by the real golden opportunities at journey's end, sailed from flourishing Persian gulf ports, down and across the Arabian Sea to trade for gold, incense and stones. These great risk-takers formed a new class of venture-some entrepreneurs. Seals inscribed with the script of the rich and civilized Harappan culture of the Indus Valley have been found on 3rd-millennium Sumerian sites, and gold beads of a type made at Ur have been found in northwest India, evidence of early contact.

Warfare brought spoils of battle. Assyrian monarchs were often content with sending their armies on treasure-hunting expeditions rather than military conquests, which brought new areas under their control. It was easier to strip a land of its gold than to police it. Assyrian warriors ranged from Cilicia in Asia Minor east to the Zagros Mountains of Iran and from Lake Van in Armenia south to the head of the Persian Gulf in search of golden stores of treasure.

Occasionally an exceptionally able ruler managed to forge a working empire out of those lands. The brutal Sennacherib, king of Assyria in the early 7th century B.C., was such a man. He took more than 200,000 prisoners from rebellious Palestine and captured almost 50 fortified cities. Jerusalem, capital of Judah, the southern part of the kingdom of Israel, was spared only because King Hezekiah paid a tremendous ransom. Sennacherib's account tallies with that of the Bible. In addition to the former annual tribute Jerusalem paid to Sennacherib, Hezekiah promised: "further tribute and presents due my majesty...He sent after me to Nineveh, my royal city, 30 talents of gold, 800 talents of silver, jewels, antimony... couches of ivory...all kinds of valuable treasures and his daughters, his harem, the male and female singers."

Sennacherib, like some of his predecessors, was paradoxically devoted both to destruction and creation. With one hand he could devastate Babylon, the jewel of cities, and with the other make Nineveh's development his greatest project. He razed Babylon, slaughtering all its men, women and children. The city was sacked and burned and then Sennacherib flooded the charred ruins with the Euphrates River channeled through the center of the city in specially dug canals so that, as he said: "In days to come the site of that city and its temples and its gods might not be recognized."

The savage emperor devoted equal fervor to the embellishment of Nineveh, founded by his father Sargon II. He and his heir, Esarhaddon, made their capital the most beautiful city in the world, filling its well planned temples and palaces with gold, silver, gems, splendid fabric hangings, ivory and rare woods. Esarhaddon, who helped murder his father, was an invincible warrior and skilled diplomat, who had more in mind than the taking of accumulated treasures when he overwhelmed the enfeebled kingdom of Egypt. His greatest endeavor was the rebuilding of Babylon, which his father had effaced from the earth. He filled his cities with treasures of every description; his parks and botanical gardens with plants and animals from three continents. His successor, Assurbanipal, inherited an empire stretching from the Caucasus gold-filled mountains in Asia Minor to the Nile.

Tribute flowed in from the far reaches, but the mighty empire was too extensive to protect its vulnerable flanks. After Assurbanipal's death in 626 B.C., the vassal lands, which had been raped and terrorized for so long, rebelled against the brutal Assyrians.

By the end of the century the once mammoth empire was fatally wounded by a coalition of Medes from the plateaus of Iran and Chaldeans from Babylonia. They laid waste to Nineveh and carted off her fabled gold just as Assyrian warriors had dome to countless cities during their centuries of power. When Layard unearthed bas reliefs depicting scenes of battle, court ceremony, and religious observance at Nineveh, he marveled at how the world had changed in the 2,500 years they had lain in darkness.

"The luxury and civilization of a mighty nation had given place to the wretchedness and ignorance of a few half-barbarous tribes. The wealth of temples, and the riches of great cities, had been succeeded by ruins and shapeless heaps of earth. Above the spacious halls in which they stood, the plow had passed and the corn now waved."

Conquest brought gold in tribute to Mesopotamia but it was the traders and long-distance merchants bearing products of Mesopotamian agriculture who kept the river cities aglow with gold and steeped in luxuries from the far corners of the known world. The bazaars offered a tantalizing assortment of manufactures, precious metals, pigments, textiles, perfumes, incense, rare woods, spices and novelties. People gathered to admire these exotic wares and listen to tales of far off lands.

The lion's share of gold came from Egypt, Arabia and Ethiopia which had exported gold from remote times. Metal from central Arabia was brought overland in donkey caravans, organized by traders who banded together for safety on the grueling desert crossing. Gold from deposits along the Red Sea, south of the Gulf of Aqaba, was most likely brought by sailing ships like those modeled in silver and copper from the Ur burials. And gold from the region of Saba, linked in legend with the Queen of Sheba, was probably embarked from the region above Aden on boats that sailed along the Arabian coast to the lower Euphrates River.

It was thanks to gold, at least in part, that men strove to develop better means of transport. Over countless centuries the tendrils of the first localized trading routes proliferated. As foot and donkey traffic was supplemented by camel caravans and wheeled vehicles, trading networks stretched farther and farther. The earliest long distance trading moved on rivers. Reed boats evolved from rafts of the 4th millennium B.C. were paddled along familiar river highways before giving way to Egyptian craft with linen sails by about 3000 B.C. Soon after, cedar ships were tentatively exploring the Mediterranean, the Red Sea and the Persian Gulf. Sailors took advantage of seasonal winds and dependable currents.

Land routes were carved out by profit-hungry merchants willing to take tremendous risks as they piloted their caravans across sandy wastes and wind-whipped mountain passes. Mesopotamians depended heavily on the Syrians, perhaps the greatest traders of all time, to act as middlemen in much of their trade, particularly the lively gold commerce with Egypt and the coasts of Ethiopia and Somalia. But they themselves ventured north from the broad river plains of Mesopotamia over the towering Zagros range's peaks to Lake Van in the mountains of modern eastern Turkey where they traded for gold from the Caucasus. Archaeologists have discovered the remnant of a colony of Assyrian merchants at the foot of a huge mound in Central Turkey, which marks a Bronze Age city. The foreign traders set up a trading organization to control and stimulate trade between Anatolia and Mesopotamia. Assyrian cuneiform tablets found there that record their activities are the oldest extant accounts of Turkish history.

The most enduring symbol of Mesopotamian might and majesty is the famed Tower of Babel, which was the most celebrated wonder of antiquity. This ziggurat was part of a great Mesopotamian temple complex that was dedicated to Bel-Marduk, the chief god of Babylon. The original temple dated back to the first Babylonian dynasty in the 19th century B.C. In the 18th Century B.C. under Hammurabi, who is best remembered for his Code of Laws, Babylon became a glittering center of power and culture, which attracted a large part of the ancient world's gold production for about a hundred years. Deciphered records tell how Babylonian strength was weakened after Hammurabi's death. Babylon was plundered and much gold carried off. The great gold statue of Marduk was taken by the Elamits from Iran. Eventually, it was recovered and returned to the temple, minus its gem encrustation.

Through the ages opposing forces vied for Babylon, destroying, rebuilding and replacing palaces and temples. In the Neo-Babylonian Empire the great Nebuchadnezzar fashioned a city that hailed itself "supreme in all the world". In the 6th century B.C. Nebuchadnezzar destroyed the city of Jerusalem, burned Solomon's temple, looted its fabled gold and exiled the Jews to Babylon to labor in one of the most ambitious building projects ever carried out. It was during this renaissance that the great temple and the Tower of Babel were celebrated as the world's most splendid monuments.

Rubble mounds covering some 500 acres along the Euphrates' east bank are all that remain of the city's plan of public buildings, parks, fortifications and temples. Cuneiform records and the description of Herodotus, who visited Babylon in the 5th century B.C., provide most of what we know about Nebuchadnezzar's city. A contemporary inscription figured the number of religious sites. There were 53 temples to the chief gods, 55 shrines dedicated to Bel-Marduk, 300 to the earth gods, 600 to the celestial divinities, 180 altars to the goddess Ishtar and 192 altars to various other divinities. All of these were under the administration of priests and priestesses who assured worshippers that the omnipotent gods showed most favor to those who made offerings of gold.

No temple attracted such rich offerings as the Esagila, the temple connected with the Tower of Babel. The exterior of the temple was plain, made of dried mud brick with gleaming doors of gilded bronze. In the dim interior there was a profusion of gold idols and furnishings. Statues of pure gold or gilded figures were dressed in richly embroidered clothes. Some were woven with golden threads and others were covered with gold plaques in a variety of shapes or gold beads and semi-precious stones. Canopies and hangings, plates, lustral bowls, thrones and scepters for the gods were fashioned of gold. The fame of the great canopy called the "golden sky" was spread far beyond the frontiers of Babylonia.

The Tower was across the street from the Esagila in a walled courtyard a quarter of a mile square. The ziggurat's architects sought to fulfill the king's command that they "raise the top of the tower that it might rival heaven." It rose 300 feet into the air from a base which measured 300 feet on each side. Each of the seven stories of the stepped pyramid was of a different color of gleaming glazed brick. The shades ranged from black to gold corresponding to the colors assigned the planets by Mesopotamian astrologers.

Atop the ziggurat was a golden chamber prepared for Bel-Marduk, originally the sun god, to rest when he came to visit earth. It was furnished with couches, tables and stools of pure gold and kept ready by a specially selected "wife" who waited for the god's descent. Herodotus described a golden idol at the base of the ziggurat that showed Bel-Marduk in half human, half animal form seated on a gold throne with his feet resting on a golden footstool. The Greek historian reported that an estimated 26 tons of gold had been used to make the idol and his appointments. As much gold, or more, was said to have gone into the chamber on top. Outside in the courtyard were enormous altars for sacrifices. One was of gold and near it was a standing figure of Bel-Marduk made of gold. It measured 12 cubits or about 20 feet high.

Herodotus saw none of these splendors for himself but most of his account has been borne out by Babylonian records unearthed in the 20th century after the German archaeologist, Robert Koldewey, first dug in the mounds to reveal parts of the immense portals and towers of brilliantly glazed brick that once guarded Nebuchadnezzar's city. Nabonidus, the weak and corrupt successor of the great king, gave up Babylon to the brilliant Persian conqueror, Cyrus the Great in 539 B.C. Cyrus and later Darius, whose empire spread from India to Lydia in Asia Minor, respected the sanctity of the Esagila and the Tower of Babylon. But Xerxes, Darius' son, was eager for the sacred gold and ordered the priests killed and the removal of all the vestments, idols, furnishings and offerings stored in the temple treasury. He then let the complex and the city itself to fall into disrepair.

When Alexander the Great occupied Babylon in 331 B.C. the city was a collection of ruins with little hint of its former magnificence. But the city's past captured the world conqueror's imagination. He envisioned his capital in a restored Babylon whose tower, once rebuilt, would symbolize his achievement. The Tower had fallen down over the centuries and many of its enameled bricks had been hauled away. But still so many lay heaped about the base that a calculation of how long it would take to clear the site convinced Alexander to abandon his grandiose scheme. His lieutenants figured it would take 10,000 men at least two full months simply to remove the rubble so the young general from Macedonia reluctantly left the remains of Babylon to the elements.

Some 4,500 years ago the king of Lagash formulated a law which read "Pay thou with good money." Gold was not money as we think of it today in that ancient time, but it played an important role in the evolution of barter into more sophisticated systems of commerce. Gold couldn't be altered, could be easily divided without damage, and maintained its value. Barter had many disadvantages for long range trading. As long as people were trading near home and everyone agreed on the value of goods and services to be exchanged, all went well. But if a man needed a copper knife but had only chickens to trade and couldn't find a metal smith who wanted them, he was stymied. Or if someone wanted a jug of beer but had only a cow to offer, he faced the problem of how to divide his valuable beast for the cheaper brew.

Values were arbitrary and the need for a standard led to the emergence of certain objects in terms of which goods and services could be measured. Perhaps the first of these were flint weapons such as those found at a Stone Age site in Russia where hundreds of thousands had been manufactured. Cattle, which were valued everywhere, replaced tools and weapons as the prevailing medium of exchange in remote times. But cattle presented certain problems. They couldn't be divided into smaller units; they varied in size and quality they had to be fed, watered and sheltered and were difficult to transport long distances.

Menes, the Upper Egyptian chief who united the river settlements along the Nile ca. 3200 B.C. to become the first of Egypt's pharaohs, minted the earliest known gold currency. He issued gold bars weighing a uniform 124 grams that bore his stamp. He also fixed a ratio of 2½ to 1 for silver in relation to gold. The ratio varied according to relative availability of the metals. Silver was highly prized in Egypt because there was none. Mesopotamians also liked silver but not as much. The ratio in early times was held at 8:1. During the reign of Hammurabi, who set fixed wages and made laws governing sales and loans, the ratio shot up to 16:1 before stabilizing at 10:1. But during the Neo-Babylonian era and the following Persian period it varied between 12:1 and 13:1.

Menes' little bars and the gold rings the pharaohs had made were too valuable for ordinary commerce. They were useful only to the landed aristocracy, not to the masses of peasants who relied wholly on barter. But in Mesopotamia, as early as the 3rd millennium B.C., small pellets of gold called shekels were weighed on scales in the bazaar and used in exchange. The shekel was both a fixed weight and a standard of exchange. The weight and value of shekels varied a great deal. The Hebrew shekel of gold weighed about 16.4 grams; the shekel in Mesopotamia weighed only about 8.34 grams. The Bible mentions "the shekel of the standard" recognized by merchants. This may have been a weight stamped with its value as opposed to an unmarked shekel which had to be weighed to determine its value. Larger units, the mina and talent, were established as multiples of the shekel as trading became more complicated.

The establishment of accurate weights benefited the goldsmith. He knew gold varied in purity and by weighing a mass of gold before and after it passed through the refining fire he could determine its quality. A clay tablet found in an Assyrian site reads: "X minas gold put into furnace and after heating Y minas gold remained; the loss as result of heating equals X minus Y minas." Once a gold lump had passed through the furnace the goldsmith knew it was as pure as could be.

Gold gradually became the one substance that was never refused in payment. Both those who had an abundance of it, and those who had no native gold valued gold as the universal store of value. It was neither so scarce as to be completely unavailable nor so plentiful as to be worthless. It was imperishable, unlike food; easily portable, unlike building materials; and divisible, unlike cattle. In addition, a very small amount represented concentrated wealth. Gold, linked with man's earliest gods, was acceptable as a substitute for human sacrifice on the world's first altars. If one man killed another, gold was accepted in payment of the blood price. The amount varied with circumstance and class. In some communities if a man murdered a slave, a small gold ring sufficed to pay for his crime. If a peasant killed an aristocrat, the amount of gold exacted was so impossibly high he had no choice but to offer his own life in payment.

Traders in the marketplace collected bits of gold as payment from mercenary soldiers, priests, fellow merchants and non-agricultural specialists who had no goods or services to exchange. This gold was taken to the goldsmith who melted it down for the trader and cast it into any shape he desired – beads or rings, which could be worn about the neck or waist, ingots, or even hooks which could be fastened to clothing.

The lumps of gold and the gold-silver alloy, electrum, so common in the mountains and alluvial sands of Asia Minor, which found their way into the marketplace, had a drawback. Because the weight and purity of each bit varied, each exchange required the weighing and assaying of the gold. The process most commonly used to determine the quality of the metal was testing it on the touchstone. This was a black, jasper-like stone also called Lydian stone, on which the goldsmith rubbed the gold to be tested. The mark left on the stone was then compared to a mark made on the stone by a touch needle.

A complete set of needles included three sets each containing 24 needles. The needles of the first set were made of varying proportions of gold and silver. Those of the second set were of gold and copper mixtures, and those of the third set of the three metals combined. The first needle was composed of 23 parts inferior metal and one part gold. The amount of gold in each needle increased, and the 24th, and last needle of each set was of pure gold. It was from this ancient system that we derving rating the gold purity rating in carats—with 24 carats signifying 100 percent purity.

Pliny wrote that "this method is so accurate that they do not mistake it to a scruple." However, it was also tedious and time consuming. For example, the pale electrum bars and lumps, which were traded among merchants by the 11th century B.C. throughout Asia Minor, varied in their gold content from 36 per cent to 55 per cent, and always had to be weighed. Lydian merchants avoided testing by stamping those lumps that had already been assayed. This was the first step in the introduction of modern coinage, although gold cast in the form of small ox heads stamped by the Minoan mint on Crete were issued long before this and served simultaneously as weights and currency.

The Lydians were responsible for the significant introduction of electrum coins of uniform weight and size. The Lydian empire derived its wealth from the golden sands of the Pactolus, which flowed past Sardis, and from international trade. In 1st-millennium B.C. the Lydian empire reached from Asia Minor's western coast across modern Turkey and grew rich as the agent dealing between the eastern nations and the Greek colonies of the Aegean coast. A network of highways converged at the splendid capital of Sardis where a beguiling assortment of luxuries collected from Assyria, Persia, Egypt and the Orient. Lydia, home of King Croesus, was so immensely rich that the king was able to give almost four tons of gold bullion and ornaments to the shrine of Apollo at Delphi.

In the 7th century B.C., about a hundred years before Croesus' reign, a king named Ardys seems to have allowed private merchants to strike their own coins. Archaeologists at Sardis have found small round pieces of electrum with deep punch marks impressed on one side. There are very few of them and they probably did not enjoy wide circulation. They appear to have been made by dropping blobs of molten electrum on a slab and hammering a punch into the metal as it hardened.

Under King Alyattes, Croesus' predecessor, official coins were first issued. They were electrum lumps marked with a lion, the symbol of royal authority, and the weight of the coin. The heaviest weighed 168 grains and was called a *stater*, or standard of value. Coins of diminishing sizes and weights represented fractional stater values. The one-third stater bore a lion's head design, and an even more diminutive coin was stamped with a tiny lion's paw.

The development of Lydian coinage culminated during Croesus' reign. He established a bi-metallic coinage in gold and silver, which was issued in such large quantities that it soon drove the electrum coins of uncertain gold content out of circulation. His gold coins were 98 percent pure. They were made from electrum refined at the royal mint at Sardis. Archaeologists have dug out more than 300 ceramic vessels in which the metal was separated into gold and silver. The lozenge-shaped staters were thick and clumsy looking, but because they bore the guarantee of the mighty Lydian king, they were acceptable to both cosmopolites and barbarians on the Eurasian frontier. Croesus set the ratio at one gold coin to 13.33 of silver, and he exercised strict control over the washing and mining of gold in his kingdom to ensure the fixed relationship of the two metals. Herodotus, who reached Sardis as well as Babylon in his extensive travels over most of the ancient world in the 5th century B.C., wrote that the. Lydians "were the first men whom we know to have struck and used gold and silver coins."

The Lydian invention of coinage played a vital part in the spread of Mediterranean civilization. With the introduction of gold coins, the royal metal became potentially available to every man, no matter how low his station in life. Ambitious soldiers, craftsmen and merchants were rewarded with gold coins that could be held without losing quality or value. Gold and silver coinage streamlined commerce. The Greek colonies struck their own issues; each city-state choosing designs symbolic of the place of manufacture. Samos' coins bore the scalp of a lion; Athens, an owl; and those of

Ephesus, a bee. Wealthy merchants were restricted from minting their own coins as they had in the past. Greek colonies based their currencies on gold because they dealt with the Persians and other southwestern Asia people who preferred gold. But the mainland Greeks, who had very little native gold, used silver from the rich Laurion mines near Athens, although gold was mined in the European colonies of Syracuse and Tarentum in the 4th century B.C.

Europe's first large scale gold issue came toward the mid-4th century B.C. when Philip of Macedonia, Alexander the Great's father, minted an immense volume of magnificent gold staters bearing his portrait. The 23-carat gold for his staters, a far cry from the rough coins of Croesus, came chiefly from the northern Greek mines of Thrace.

Following the Roman conquest of Macedonia in the 2nd century B.C. great numbers of Philip's coins were circulated in Rome where they served as that city's first gold currency. They were used all over Europe and Asia. "Philippus" was the generic term for any fine gold coin. Barbarian nomads who moved westward across Europe and into the British Isles used the staters as the model for their first coins, transforming the superb Greek designs into primitive but striking images of horses and men.

Coinage created the need for moneychangers and expanded banking facilities. Temple treasuries continued to serve as repositories for royal, civil and religious treasure. But as more gold came into private hands, the goldsmith found himself not only melting coins to fashion jewelry and offerings into current tastes, but also acting as a banker. He was well suited to such a function since he was familiar with the metal, had facilities for storing it and could transform it from one shape to another.

Gradually the goldsmith expanded his banking activities to include drawing up contracts and making loans. He was the first private insurer, loaning out at interest part of the gold entrusted to his safekeeping. He gave his depositors a receipt for the gold and soon a system of bills of exchange evolved. The precious metal backing the receipts and bills of exchange piled up in his vaults.

Sometimes, he lent too much and was unable to make good on paper presented for exchange. To counteract this and other such abuses as outrageous interest charges, laws governing banking were passed in most countries. In many places it was a crime to charge any interest at all. However, despite the regulations money-lenders continued to charge interest and never lacked for clients.

Goldsmiths acted as bankers throughout history. In European cities goldsmiths' districts often became banking centers. Lombard Street, the financial heart of London, was once goldsmiths' row. The founder of the Bank of England and many other early English bankers started their careers as "goldsmiths who keep running cashes." In the Late Middle Ages and the Renaissance, whole dynasties of Italian and German international banking families were responsible for funding wars and New World exploration. They propped up shaky regimes, bailed out bankrupt sovereigns and were sometimes more powerful than the kings they served.

The Mesopotamian traders who used bits of weighed gold to ease the chafing limitations of barter had no way of foreseeing how coinage would develop. Even Croesus had no idea of the extent to which gold- and silver-standardized currency would stimulate human activity. The fabulous wealth of Croesus' empire attracted the attention of the Persians who were growing stronger under the military genius of Cyrus the Great. Croesus sent to the Oracle at Delphi to ask if he should challenge the Persians. Replying with characteristic ambiguity, the Oracle prophesied that if the Lydian king attacked the Persians, he would destroy a mighty empire.

The king was jubilant at what he interpreted as a promise of victory. He sent magnificent thanksgiving gifts to the shrine of Apollo. They outshone the royal gold throne Midas, the first non-Greek to make an offering at Delphi, had sent from the kingdom of Phrygia, east of Lydia. They even outdid the gifts sent by the earlier Lydian king, Gyges, who, according to Herodotus, had offered "a vast number of vessels of gold, among which the most worthy of mention are the bowls, six in number, and weighing altogether 1,800 pounds, which stand in the Corinthian treasury."

Croesus sent two gold staters to every citizen of Delphi, couches plated with gold and silver, gold vessels and ceremonial weapons, large golden statues, jewelry and an immense treasure in gold bullion. The enormity of it all must have stunned the Greeks who seldom saw gold in any quantity. Unfortunately, Croesus had been too hasty. He made war on Persia and a mighty empire was indeed destroyed, but it was his own.

Sardis, and with it Lydian might, fell to Cyrus who went on to mastery of the ancient world. He adapted the Lydian invention of coinage, and within a half century the Persian bi-metallic system had spread throughout Asia Minor and become a vital force in commerce, taxation and everyday life. The Lydians introduced true coinage but the Persians were the first to base the state economy on money. Coinage had come a long way from the first crude bean shaped lumps stamped by merchants to the dazzling Darics, which spread the portrait of the great Darius from the amber shores of the Baltic to Africa.

In expediting trade, coinage fostered the growth of communications among previously isolated areas. Money served as capital and encouraged the industrial growth. The man who had previously reckoned his wealth in cattle or cumbersome cast ox hides of bronze was able to make advantageous transactions in distant places. A new feeling of security and self-confidence was enjoyed by those who could accumulate gold and silver coins, choosing whether spend them or accumulate them against later need. Even slaves were sometimes able to save enough to buy their freedom.

Coinage helped foster the development of the arts from goldsmithing to poetry, which was often commissioned by wealthy patrons who paid in gold for verses praising them. Coinage quickened the tempo of life and added to its growing pleasures. For more than 2,500 years coins, some roughly made or debased and others representing masterpieces of the engraver's art, have been a vital part of history, linking one era with another and providing a gallery whose images reflect the ebb and flow of human events.

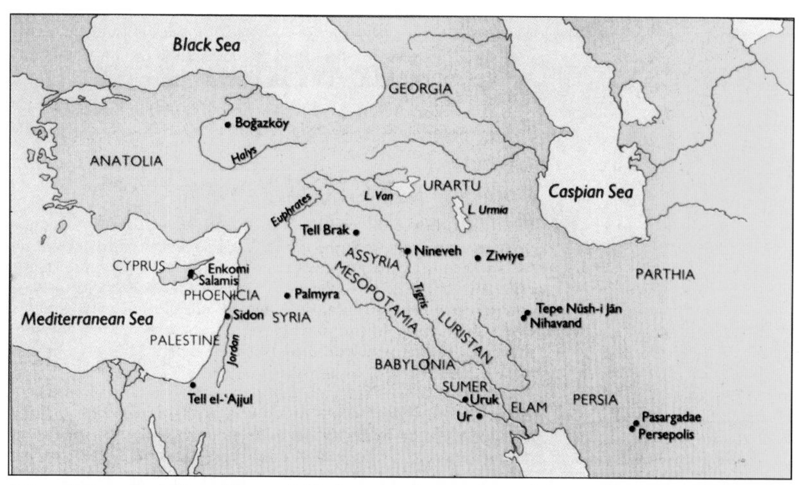

(Above) Near East Map. *(Below, left):* Assyrian bronze bust of Sargon II who boasted in his annals that seven Greek kings of Cyprus paid him gold in tribute. ca. 9th century B.C. *(Below, right):* Babylonian king receiving golden tribute ca. 2000 B.C.

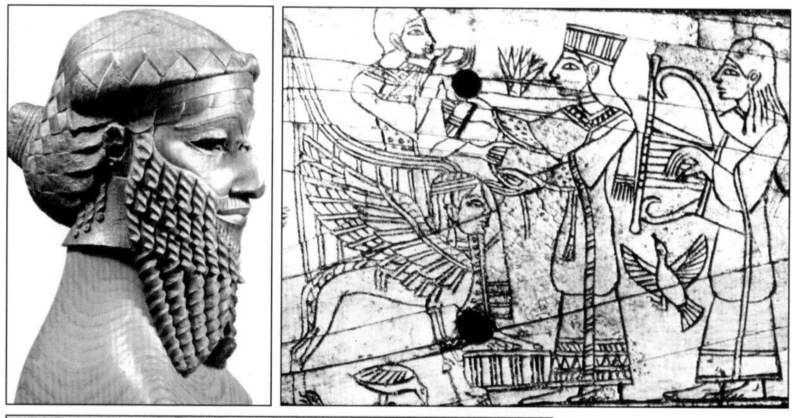

(Left) Painting depicting Assyrian city ca. 9th to 8th century B.C.

(Left) Built about 575 BC by Nebuchadnezzar II, the Ishtar Gate was one of the eight gates of the inner city of Babylon, capital of the Babylonian Empire (now in Iraq). It is one of the most impressive monuments rediscovered in the ancient Near East.

(Right) Achaemenid gold bracelet with horned griffins from Oxus Treasure 5th century B.C.

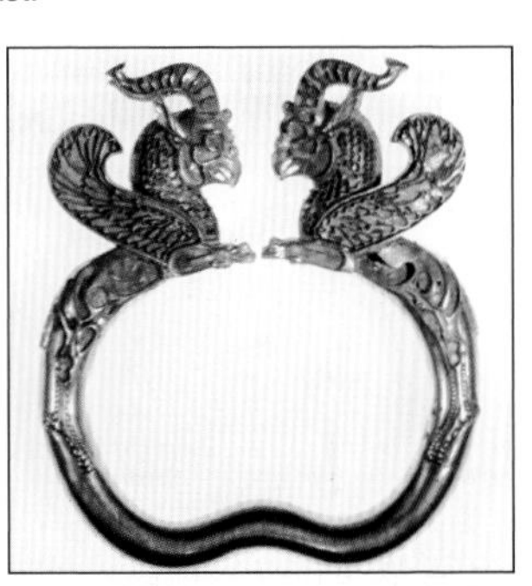

(Below) Urartian tombs from late 8th century B.C., located in present-day Turkey.

CHAPTER 4

GOLD OF THE PHARAOHS

The bleak Nubian desert of the northern Sudan is dotted with great mounds of tailings. More than a hundred hills of sand and crushed rock mark the location of gold mines where many thousands of years ago slaves produced the shining metal for Egypt's pharaohs. The monotony of this treeless landscape is occasionally relieved by the remains of round stone huts. In them are ancient querns, the granite hand-mills used to crush gold-bearing quartz. Still standing above galleries which twist and turn for miles beneath the sands are the ruins of the tilted stone tables where the powdered ore was washed.

These memorials to an ancient hell where countless thousands died of hunger, thirst and exposure, contrast sharply with the breathtaking beauty of Tutankhamen's golden treasures. The dramatic discovery of the young king's gold-filled tomb less than a century ago stunned the world. Tutankhamen's 14th century B.C. tomb was the only one in the Valley of the Kings on the west bank of the Nile at Luxor to have escaped the notice of tomb robbers who had plundered royal resting places over the centuries.

The astounding treasure-laden tomb found by Lord Carnarvon and archaeologist Howard Carter is familiar to people all over the world. In fact, the overwhelming richness of his funerary equipment has made the historically insignificant boy king the world's best known pharaoh and a symbol of the golden magnificence of ancient Egypt. But there is more to the story. Tutankhamen comes into it rather late, more than 2,000 years after early Egyptians had begun to wash alluvial gold from the Nile.

The Egyptians were the greatest producers and consumers of gold in antiquity. Gold and Egypt have been as closely linked as Egypt and the all-important Nile, which supplied the world's first great gold harvest. The Egyptians were the first to extract gold from quartz reefs. They not only had vast deposits of the sacred metal but, just as important, they had the political organization which enabled the strong central government to draft labor for the service of the state. It was Egypt's incredible reservoir of manpower which built the pyramids and also gave Egypt supremacy in gold production.

Egyptians worshipped the sun and their pharaohs, who were considered divine sons of the sun. Gold for them was symbolic of the sun and its use reserved for religious purposes. In theory, gold was the rightful property of the pharaoh and his fellow gods. It was used lavishly to display the might and majesty of the awesome deities. In practice, however, some gold found its way into the hands of others. Especially in very remote times, there was probably little control over random gold washing from the Nile. Later strict government control, especially over the mines, made it difficult, if not impossible, for individuals to gather gold. Yet they found another way to get the coveted metal. Tomb robbers, active since earliest times, rifled new graves and old, stripping them of treasure given the wealthy dead to ease their journey to the underworld. Much of the superbly crafted gold ornaments and furnishings were melted down and found their way into barter markets. At times such gold was fairly easy to attain that a young married couple could acquire a little bit of gold as a sankh, or nest egg.

No other nation came even close to amassing as much gold. There are no figures on total Egyptian production but it has been estimated that by the time of the heretic pharaoh Akhenaten, thought by some to be the father of Tutankhamen, more than a million and a half pounds of gold had been produced. Gold was one of Egypt's earliest exports. The fame of her gold spread so far and wide that foreign rulers continually begged for it. During Akhenaten's rein in the first half of the 14th century B.C., the king of Mitanni, a land north of Mesopotamia, wrote: "Send me much gold, more gold; for in thy land gold is as common as dust." One king of Babylon complained of a gift of 20 minas of gold which was reduced to five when refined. Another Babylonian king grumbled when he received only two minas. "My brother has sent me as a gift but two minas," he dictated to a scribe who wrote on a clay tablet. "If now gold is abundant with you, send me as much as your fathers did. But if it is scarce, send me at least half as much."

The gold of the pharaohs was very much on the minds of gold seekers more than 3,000 years later. At the end of the 19th century, when the shining promise of the California gold strikes had faded and the bonanzas of Australia, Alaska and the Klondike been exhausted, a number of hardy prospectors turned their attention to the long-abandoned mines in the Nubian desert.

They were convinced the mountains of tailings were rich in gold. After all, they reasoned, the ancients used only primitive methods which couldn't possibly have extracted all the precious metal. Much to their amazement these modern miners found that the ancient extraction process had been so efficient that almost no trace of gold remained.

Gold seekers had better luck in the wild Eastern Desert of Upper Egypt. At the turn of the century several English mining groups reopened some of these deep mines which had been the first in the world. A large percentage of ancient Egyptian gold came from the plateau of gold-bearing quartz bordering the Red Sea and about 60 miles wide. It slopes westward from forbidding, waterless mountains to the Nile.

The mines, scattered from Philoteras southward to Berenice, are still visible although their entrances have been blocked by drifting sands. Some of the ancient shafts are over 300 feet deep. Galleries hewn out of solid rock wind up to six miles underground, pursuing the veins of gold. Despite millennia of exploitation, the 20th-century prospectors were not disappointed. They worked 20 years and brought out three million dollars worth of gold plus a staggering number of human bones—the remains of slaves whose endless labors furnished the dazzling splendors of pharaonic Egypt.

Nub was the ancient word for gold and Nubia was the land of gold where some of the very earliest Egyptian gold artifacts come from. Shallow graves scooped in the sands of Lower Nubia have yielded a few clumsily made amulets at least 6,000 years old. They were made of alluvial gold washed from the Nile sands where it had been collected by rock-scouring currents since the beginning of time. The alluvial or placer deposits in Egypt, Nubia, the Sudan and Ethiopia were the first to be worked because it was relatively simple to pan for gold with a shallow basket or calabash. At a somewhat later period alluvial gold, deposited just beneath the surface of the earth in remote geological ages on higher terraced land flanking the Nile, was mined in shallow trenches.

The rich placer deposits from the first to the fifth Nile cataract along with production from the Eastern Desert accounted for most of early Egypt's gold. Expansionist pharaohs later enlarged their empire to include the Nubian desert and the area south of the fifth cataract on the Ethiopian Plateau where government prospecting expeditions found abundant new sources.

Nature showered the Nile kingdom with seemingly inexhaustible amounts of the metal, which was believed to be the essence or body of the sun. But the appetite for gold feeds itself, and throughout history there has never been such a thing as too much gold. During the zenith of Egypt's history in the New Kingdom when the country became a great military power, a stream of gold inundated the imposing stone temples and palaces raised by the pharaohs. It came as raw bullion and manufactured articles of exquisite workmanship from scores of foreign cities. Some was booty from looted lands; some was tribute rendered by vassal states or sent as diplomatic gifts by potential victims of Egyptian might and some came in trade.

Thutmose III, perhaps the greatest of all pharaohs, ruled over 400,000 square miles from Syria to the Sudan. He was a great conqueror, a just ruler and a lover of beautiful things. He designed the magnificent sacred vessels for the sanctuary of Amon and adorned many temples with the immense golden spoils that fell to him. His annals list 17 separate campaigns against Syria and Palestine. A scribe recorded some of the plunder taken in one of them. The prizes included flat dishes of costly stone and gold, gold knives, "a staff of carob wood wrought with gold and all costly stones in the fashion of a scepter, the head of which was inlaid with lapis lazuli and much rich clothing of that foe." Thutmose was so impressed with the skill of Levantine craftsmen that he imported Syrian goldsmiths to teach their secrets to the Egyptians.

It was his grandfather, Thutmose I, who had established Egyptian control over Gush which comprised Nubia and the Lower Sudan. He had marched his armies as far east as the Euphrates, stripping Mesopotamian temples, palaces and ziggurats of their gold. He had so much gold that some of it was exported to Crete, Syria, Mesopotamia and mainland Greece, along with Egyptian mining and goldworking techniques. Thutmose I initiated the monumental additions to the temple of the Theban Amon at Karnak, which became the world's largest and most splendid. Thebes became the richest city on earth, a center for international trade. Gold filled the palaces of the pharaohs, the hundreds of temples and the mansions of the aristocracy with every sort of exotic luxury. Gold bought timbers for construction and shipbuilding, purple dyes, precious amber and silver for a kingdom that had none of these.

Gold financed monumental building schemes and furnished superb raw material for the most gorgeous works of art. Amenhotep III, the grandson of Thutmose III, used the gold which poured into the royal coffers in tribute and trade to pay for splendid temples including the famed temple of Luxor and imposing temples in Nubia. During his prosperous and relatively peaceful reign he also used gold to maintain friendships with the powerful kings of Cyprus, the copper island, Babylonia, Mitanni and Assyria. A stele inscription proclaims the wonders of the temple he erected at Thebes: "an eternal, everlasting fortress of fine white sandstone, wrought with gold throughout; its floors adorned with silver, all its portals with electrum...It is supplied with a 'Station of the King' wrought with gold and many costly stones. Flagstaffs are set up before it; it resembles the horizon in heaven when Re (the sun god) rises therein. Its lake is filled with the great Nile, lord of fish and fowl."

Herodotus called Egypt the "gift of the Nile." The colossal river system, which runs over 4,000 miles, was all important. Without the river the Egyptians would have struggled in a barren land where no rain falls. Because of the great river that "creates all good, issues from the earth and comes to keep Egypt alive," a great civilization flowered in the desert for 3,000 years. About ten thousand years ago small settlements of proto-farmers lived along the verdant Nile banks. They depended on its annual rhythm of flood and ebb to fertilize and irrigate the barley which ripened under the warming eye of the sun.

They made observations of seasonal changes. Working together they made canals, erected dikes, sunk wells and made catch basins to hold the receding flood waters. They also gathered alluvial gold in a haphazard way and used it to fashion offerings to the sun god or for rare ornaments such as the 6,000-year-old cold-hammered pendants and bracelets found at El-Gerza, south of modern Cairo. Real gold collecting didn't begin until the scattered communities, that had been divided into the two kingdoms of Upper and Lower Egypt, were united by the tribal leader of Upper Egypt. Narmer, called Menes by the Greeks, established the capitol of his kingdom at Memphis. There, at the meeting ground between Africa-oriented Upper Egypt and Western Asia-oriented Lower Egypt, Menes' successors for the next four centuries built their gold-filled tombs and pyramids.

The fact that Menes issued standardized gold ingots indicates that he controlled production and could mobilize the labor to extract and refine gold found in the Nile, on the terraces and deep beneath the harsh desert slopes. In the 4th millennium gold production was already organized into a large-scale industry, which was also a royal monopoly. Even a river nugget found by a fisherman belonged to the pharaoh. The nature of the government in which all authority was vested in a ruler who was so sacred that his name could not be spoken, and the Egyptian genius for organization, made possible the large-scale exploitation of distant gold sources. The ultraconservative shape of Egyptian society crystallized under the early god-kings, called pharaohs after their dwellings.

In antiquity there were fewer people competing for luxuries. There were relatively few precious substances and gold was by far the most prized. When gold was concentrated in the hands of a ruler who was an earthly god, it could be used to awe the masses and enhance the supposed link between heaven and earth. No rulers used gold to better advantage than the pharaohs of Egypt. During the Old Kingdom, from ca. 2700 B.C. to ca. 2200 B.C., the pharaoh was the focal point of every aspect of human activity. The wellbeing of him as the son of the Sun, an isolated and aloof being, was crucial to the fortunes of the people.

Gold was employed to make his surroundings as resplendent as befitted a deity. His daily routine was circumscribed by ritual. Gold covered his body, his regalia, furnishings and playthings. His utensils were of beaten gold, his throne, couches and tables of gold inlaid with ivory, ebony and precious stones. His arms and chests glittered. Light blazed from his great helmet crown, his broad breastplate and collars. It flashed from bracelets, rings and sandals making him shine like the sun itself. The pharaoh was carried about on a gold-sheathed litter. He rode in a golden chariot and sailed on royal barges whose several decks were embellished with sheet gold and mounted with golden fittings.

The gold of a lifetime went into the tomb of a dead pharaoh or aristocrat, supplemented by additional funerary gold. In the case of a royal tomb, the amount of gold used in specially ordered funerary equipment was colossal. The single most striking feature of Egyptian civilization was its emphasis on eternal life. In the early period only the royal family was entitled to an afterlife. But eventually nearly everyone was encouraged to look forward to existence in the netherworld. The poor were provided with the simplest supplies—clothing, perhaps a knife, a hoe and bowl. Aristocrats and wealthy bureaucrats who enjoyed the pharaoh's favor spent much of their lifetime preparing for death. They built tombs for themselves. Artists and sculptors were commissioned to depict scenes from their lives to decorate the tomb's interior. Goldsmiths fashioned exquisite articles for use in the afterlife. Cabinet makers worked in woods brought from the Levant and Africa making chests, litters, tables, canopied beds, even chariots, for entombment. Such furnishings were adorned with overlays of rich gold work, veneers of exotic woods and inlays of hippo ivory, amethyst, turquoise, jasper, garnet and multi-colored agate.

The mastaba tombs, the pyramids, the cliff tombs with all their contents, scenes and inscriptions and papyrus accounts of the elaborate mummification and funerary practices of the ancient Egyptians, are evidence of how a very small portion of the population was able to mobilize the energies of the state in preparation of a splendid afterlife. Farmers were required to donate labor. Craftsmen from leatherworkers and potters to artists and goldsmiths dedicated their sensitivity to line and style and great skill which made even utilitarian objects into works of art.

The Egyptian burial place was regarded as the house of the dead whether it was a mud brick mastaba, a pyramid, or a chambered tomb cut into limestone cliffs of the Theban hills. They were furnished to house a body, which requires food, drink, and the performance of various services. Before the Second Dynasty slaves, including musicians, singers and priests, were sacrificed to serve their lord. They were replaced by marvelous scale models in wood, clay or metal. These ushabti figures engaged in activities such as brewing, baking, fishing or making jewelry were activated by magic charms. Finally, paintings and bas reliefs with inscriptions supplanted the models. Together these things have provided the best source of information about life in prehistoric Egypt.

The history of tomb design was greatly influenced by the contest between gold and the thief. No culture ever buried more gold with their dead than the Egyptians for whom continued life was as real as tomorrow is for modern man. Ideally, the gold immobilized in the dry darkness of the tomb was eternally safe. However, this was seldom the case. Tomb robbers were quick to violate the sanctity of the tomb. By the time Thutmose I began to prepare for his tomb in the 16th century B.C., almost all the royal pyramids which had been built over the preceding 1,200 years had been looted of their gold. Thutmose decided to prepare a hidden tomb. He had the first of the cliff tombs hacked out of the rosy limestone near Thebes. The slaves who made it were killed to silence them. After his death his mummy and gold were placed in the hidden tomb and the narrow entrance covered over to resemble the bluff face. One by one other galleries and secret chambers were carved in the limestone and filled with gold. But it proved almost impossible to fool the professional thieves who managed to plunder almost every tomb in what came to be known as the Valley of the Kings. Intricate webs of hidden chambers and blind tunnels might foil thieves for hundreds, even thousands of years, but in the end they succeeded. None of the guards, modeled, sculpted or brightly, painted deterred the gold-hungry robbers.

The first Egyptians were buried in depressions scooped in the sand and covered over with loose heaps of stones to keep scavenging animals away. When Menes unified Egypt, mud bricks were arranged for a sloping, bench-like tomb. From flat topped tombs with two underground chambers to elaborate structures with up to 30 chambers which rose 16 feet from the ground. In the Third Dynasty, Imhotep, the architect of the Pharaoh Zoser, designed the world's first monument of stone masonry. Built ca. 2600 B.C. it was a stepped pyramid, much like a Mesopotamian ziggurat, built of small stone blocks. Around the 200-foot-high structure Imhotep designed a cluster of beautiful buildings including two tombs of royal relatives which had slender columns carved into the limestone facades.

His step pyramid paved the way for the majestic pyramids which remain such marvels of technology. Less than 200 years after Zoser, the Great Pyramid at Gizeh was built for Khufu (Cheops). It was the first of the three towering pyramids of the Fourth Dynasty which are guarded by the monumental and inscrutable Sphinx who lies in the desert sands at the head of the Nile Valley. They are the best known of the more than 80 pyramids which have withstood the ravages of time and the predations of men eager for building stones. Nearby are a multitude of aristocratic tombs sunk in shafts, some of them going down 80 feet in a futile effort to prevent looting.

The god-king Khufu was able to muster a reported 100,000 laborers to work on his colossal death monument, and to clothe, feed and house them during the 20 years it took to complete the construction. The Great Pyramid covers 13 acres. It is built of 2,300,000 blocks of precision fitted limestone. They were cut with the simplest copper and stone tools and moved without benefit of block and tackle. The blocks weighed an average of two and a half tons. Some weighed as much as 15 tons and yet they were moved into place by legions of straining workmen who had never seen a wheel or a draft animal.

Khufu's pyramid was honeycombed with secret passages and cleverly concealed chambers designed to deceive thieves. Imhotep's ingenious plans failed to safeguard the pharaoh's gold. All that remains of his treasures is a goldsmith's weight of 200 grains stamped with his mark. It was found in a royal workshop and shows that he maintained rigid control over goldworking. By a miracle the treasures buried with Khufu's mother, Queen Hetepheres, escaped detection until they were found 4,500 years after her death by a Harvard-Boston Museum expedition.

The queen mother's tomb was partially plundered soon after her death. Her son ordered her sarcophagus and funerary equipment removed to a deep shaft tomb. When the American archaeologists chanced upon it they were delighted to find many beautiful articles which bear comparison with the art works of any age. Among Hetep-here's possessions was an elegant ebony chair covered with gold. Inlaid in the wooden slats of the back were tiny human figures, animals, birds, trees, tools and weapons of solid gold which were hieroglyphs listing the name and honorific titles of the prehistoric queen.

The earliest royal burial gold was found in 1900 by the great Egyptologist Sir Flinders Petrie. In the tomb of Zer, a First Dynasty king of the 31st century B.C., Petrie found the mummified arm of Zer's queen, which had been ripped from her bandaged mummy by robbers who thoroughly looted the tomb at some ancient date. The dismembered arm had been shoved into a hole in the wall by the thieves. The tomb was at Abydos, 300 miles south of Cairo, one of Upper Egypt's most ancient and prosperous cities and the center of the important cult of Osiris. In addition to the royal tombs, archaeologists have found many rich tombs in a cemetery for private people, which was used from the First Dynasty to Roman times.

On the queen's arm were several bracelets. Their beauty and refinement, although not comparable to the superb goldwork of later centuries, show that goldsmiths in the Fourth Millennium were the descendants of many generations of skilled craftsmen. One bracelet was formed of repoussé gold plaques in the shape of a temple façade with a hawk on top. The plaques were separated by panels of turquoise. Another bracelet was formed of delicate gold flowers and the seedpods of desert plants. Turquoise, amethyst and hollow gold beads held together with gold wire braided with ox tail hair made up a third bracelet, and the fourth mixed gold beads with beads of lapis lazuli.

Egyptian goldsmiths are shown at work in a number of wall paintings and reliefs. A relief in the tomb of Mereruka at Sakkara from the mid-third millennium, a wall painting from the tomb of a royal official, and paintings at Beni Hassan from the early second millennium depict goldsmiths weighing ingots of gold, recording gold weight, and puffing through blow pipes to increase the temperature of the fire to just over 1,000 degrees F. at which gold alloyed with copper, as most Egyptian gold was, will melt. They are shown pouring molten metal and beating gold into sheets. Hieroglyphics explain illustrations of goldsmiths overlaying necklaces, shrines and doors with sheet gold. Working in pairs, goldsmiths are shown bending crescentic gold collars and necklaces which terminate in falcon heads. A stylized necklace was the first symbol for gold, later replaced by a sun circle hieroglyph.

Egyptian goldsmiths were superlative technicians. A hieroglyphic inscription translates "No limit may be set to art, neither is there any craftsman that is fully master of his craft." The goldsmith was the king of craftsmen who honed his skills until he was the greatest technician, if not the most inventive artist. He could turn out a piece of filigree as breathtakingly delicate as a spider's web. In contrast, he could execute large bold objects whose smooth surfaces invited adornment in repoussé, chasing, enameling or inlay. The goldsmith's craft was generally hereditary as a father passed his apparently magical powers on to his sons. In the 12th Dynasty, ca. 1900 B.C. during what historians call the Golden Age of Egyptian art and craftsmanship, a master goldsmith named Mertisen vowed none would succeed in his craft "but I alone and the eldest son of my body. God has decreed to him to excel in it and I have seen the perfections of his work in every kind of precious stone, of gold and silver, of ivory and ebony."

Egypt was the first nation to export gold and goldworking techniques. But foreign smiths were captured or enticed away from Mesopotamia and the Levant to work in the shops of the temples and the pharaohs. Egyptian smiths were seized in battle too and taken abroad to spread their secrets. In the 6th century B.C. when the Persians, the second greatest gold power in the ancient world, invaded Egypt they took goldsmiths back to work in the great ceremonial capital at Persepolis.

The Nile smiths knew how to refine gold by smelting and cupellation but they often preferred to work with various alloys since the range of hues gave added richness to their work. Much native gold was mixed with copper and had a brassy or reddish sheen. Some had flakes of platinum large enough to be visible. Other gold with a purplish tinge contained enough iron to attract a magnet, and gold imported from the far off land of Punt had a greenish cast because of the presence of small amounts of silver. Some smiths coated copper objects with gold and passed them off as the genuine article. Ancient texts describe a process for gold plating that could pass the touchstone test but not the trial by fire. The alluvial gold from the Upper Nile deposits was purer, yellower than the quartz reef gold from Nubia and the Red Sea littoral, which contained more silver. Foreign gold from Libya, Cyprus, Africa and Asia gave the goldsmiths a wide range of colors to work with.

Change was not encouraged in the static Egyptian society so goldsmiths suppressed innovations. Over the millennia successive generations of smiths refined and elaborated traditional styles preferring symmetry and directness to artistic imagination. The basic motifs remained unchanged. Goldsmiths relied on the traditional stylized floral and leaf patterns, animal and bird forms, the representation of deities and their symbols and abstract and geometric designs. The directness of line and boldness of Egyptian goldwork and symbolism have great appeal to modern taste. The potent life symbol the ankh, for example, which was often fashioned in gold, is now back in vogue. People everywhere are seen wearing the cross with loop at the top.

The goldsmiths turned their hand to pieces which ranged from the miniscule to the massive. The Cairo Museum displays a bracelet made about 3300 B.C. of gold wire as fine as a hair. In 20th century B.C. goldsmiths mastered the challenge of sweating the tiniest gold spheres by the thousands onto a gold surface to make shimmering textured patterns. This technique of granulation, which was perfected by the Etruscans in the 8th century B.C., was forgotten after Roman times until rediscovered later. Solder of a copper or silver gold alloy was used to fasten spirals of gold wire to gold backing to create exquisite filigree designs of almost gossamer delicacy.

The same level of accomplishment was seen in large scale gold work. Whether they were casting gold, making holloware or beating the precious metal into expanses of filmy sheet to cover architectural elements, Egyptian goldsmiths were justly self-confident.

Only a virtuoso could have created the six-foot-long solid gold coffin of Tutankhamen which weighed almost 2,500 pounds. No one in antiquity matched the skill of the Egyptians at working in cloisonné, not even the Sumerians from whom they learned how to create the little cells which were filled with semi-precious stones or other materials in a variety of colors. An inlaid gold pectoral jewel belonging to Ramses II of the 19th Dynasty has 400 pieces of inlaid turquoise, carnelian, lapis and garnet. In the 19th century B.C. goldsmiths learned to make enamel inlays for cloisonné pieces by heat-fusing a powder of colored glass on gold. A few centuries later they perfected the difficult art of inlaying a gold design in another metal. A bronze dagger blade from a Theban tomb was inlaid by chiseling out a delicate pattern in the bronze, then hammering gold into the grooves and finally making an almost invisible bronze lip over the gold to lock the inlay in place.

The pharaoh was served by an elite of officials and priests who supervised lower echelon administrators. These trained bureaucrats were directly responsible for supervising the grandiose state projects of monument-building, mining and irrigation. They also served as tax collectors. Hereditary castes of aristocrats, priests, bureaucrats and high military officials made up the small closed upper class. Generally their lives were pleasant. They often enjoyed the pharaoh's favor, receiving landed estates or other gifts that might be withdrawn if they incurred his displeasure. The more fortunate of them basked in the pharaoh's golden glow, living in leisured luxury. Their homes were spacious buildings of sun-cured brick set in large formal gardens. The many rooms for public receptions and family living were elegantly furnished and boasted gold objects and other treasures. Servants, often slaves taken in foreign battles, performed the work which made such large, self-contained establishments run smoothly.

The superintending officials of the pharaoh sometimes found their fortunes in the gold mines. Men who enriched the king enjoyed his particular esteem. A man named Ameni was superintendent of the official "Expeditions to the Mines" about 4,000 years ago. On the walls of his tomb he had written: "I went to get gold for his Majesty, may he live forever. I went with a company of 400 men of the pick of my soldiers who, by good fortune, arrived without loss of any of their number. I brought the gold I was ordered to bring and the king's son honored me."

The road to the goldmines in Egypt's Eastern Desert is shown on one of the world's oldest maps. On papyrus sheets, it traces the transport route of a statue to the Wadi Hamrnamat in the late 2nd millennium. Hieroglyphs mark wells, trails, mountains, where gold was mined and where it was washed. Housing for government overseers is also shown. Some mining operations were so large that 2,000 huts were built for officials while the enslaved miners received neither housing nor care.

The gold lay in zones of barren waste. The Gush was a searing land of desert slopes. Forced armies of captive men, women and children were marched over the scorching sands and across the bleak mountains under the whips of government overseers who understandably hated their assigned duty. Sometimes the captives were luckless Nubians. There was no food, no water, no shade. The soldiers drove long lines of slaves and donkeys before them. If a child stumbled or lagged behind, he was left as his mother was forced ahead under the lash. Sometimes, before wells were dug along the way, as many as half of the slaves perished on their way to Egypt's Siberia. Most of the pack animals died of thirst. Ironically, they carried loads of water but were allowed to drink none of it because it was needed for washing the gold ore.

The Pharaoh Seti in the 14th century B.C. sent men to find water along the tortuous route. They dug hole after hole until water was found 200 feet deep at Redesieh. A frieze at Karnak shows the powerful Seti beating a multitude of slaves among which are Nubians of the Cush. He erected a temple at Redesieh to commemorate the improvement of the road to the mines, dedicated to Hathor, the cow-horned goddess of love and happiness who was also the protectress of the Eastern Desert. In an inscription she declares, "I have given thee the gold countries." Although the way had been made somewhat easier, the journey to the goldfields remained perilous with whole caravans sometimes hopelessly losing their way in blinding sand storms.

Survivors of the march were less fortunate than those who died on the way. The men, women and little children who labored at the mines were freed only by death, which usually came soon since they were condemned to work naked and without rest. The state found it cheaper to work the slaves to death and replace them than to supply food, water and shelter. One contemporary observer noted that thirst-crazed slaves were forbidden to even lick their hands, which were damp from washing the ore or tossing water on the subterranean fires that were set in galleries to crack the rock.

If exposure, starvation or disease didn't kill them, there was always the danger of being crushed beneath tumbling rock in the cramped galleries lit only by the flicker of oil-burning headlamps. In the 3rd century B.C. Theophrastus, a Greek natural scientist, described the manner in which ancient miners worked lying on their backs or sides in the narrow, two-foot-high galleries. Foot by foot they hacked their way along the vein face carving out miles of galleries. Fires were sometimes set and then quenched to crack stubborn rock. There was scant air in the deep dark. As the fires further depleted the supply of oxygen, miners died of suffocation or from air poisoned with arsenic released from the rock by fire.

Mining changed very little through the ages. The conditions described by visitors to the Egyptian mines, the Greek mines or the Roman mines in Spain held true for gold diggings up to modern times. The famous medieval metallurgist, Agricola, described what happened to those who inhaled arsenical oxide. In his 16th century treatise he wrote that some, the lucky ones, "swelled at once and lost all movement and feeling and died without pain." Others found their hands and feet puffing up like pig bladders. Their skin erupted in hideous and painful ulcers. In Roman times these sores were treated with a salve which contained gold. Lungs which inhaled the poisoned air rotted and eyes which came in contact with it were blinded.

Agatharchides, the Greek tutor of the Alexandrian Ptolemies in the 2nd century B.C. visited the gold mines near the Egyptian coast of the Red Sea. His account described them in detail and was incorporated in the work of Diodorus, the Sicilian-born Greek historian, who visited Egypt a half-century before Christ was born.

"On the confines of Egypt and Ethiopia is a place which has many great gold mines, where the gold is got together with much suffering and expense...the kings of Egypt collect together and consign to the gold mines those who have been condemned for crime and who have been made captive in war...sometimes only themselves but sometimes likewise their families. Those who are thus condemned to penal servitude, being very numerous, and all in fetters, are kept constantly at work both by day and night without any repose, and are jealously guarded to prevent their escape; for they are watched by companies of barbarian soldiers who speak a language different from theirs, to prevent their winning any of them over by friendly intercourse or appeals to their humanity...unkempt, untended as they are, without even a rag to hide their shame, the awful misery of these sufferers is a spectacle to move the hardest heart. None of them, whether sick or maimed or aged, not even weak women, meet with compassion or respite; all are forced by blows to work without intermission, until they expire under this hard treatment...they welcome death as a blessed change from life."

Diodorus described how the ore was gathered and prepared. Iron hammers assigned to the strongest men and fire were used to break up the rock, excavating subterranean passages in pursuit of the gold veins. Children carried rock to the mine entrance and prisoners "who are more than 30 years old" pounded the quarried stone in stone mortars with iron pestles. When the ore pieces were reduced to the size of bean pods, women and older men ground them in hand mills, two or three people assigned to each handle until the ore was ground as fine as wheat flour.

A modern analysis of the tailings piled high at one of the Red Sea mines indicates that a million tons of ore were removed from that mine alone which was worked up to the time Julius Caesar went to Egypt. When the ore had been reduced to a powder, it was rubbed on wide, slightly tilted stone tables or boards as water was poured on it to wash away the lighter earthy material while allowing the gold particles to be collected.

The gold dust was weighed and then refined by cupellation. The gold was put in clay vessels called cupels "according to a fixed weight and measure, adding in proportion to the quantity lumps of lead and salt, with a small amount of tin, and husks of barley." The cupels were then sealed with a smearing of mud and baked for five days and nights. The salt served as a fluxing agent fusing the fine grains of gold together. The tin mixed with the gold making it harder; the bran was a reducing agent helping remove those impurities which would oxidize and the lead united with those which wouldn't to form a slag.

When the cooled cupels were opened, a bright button of gold was found in each one. These were melted and usually cast in the form of rings about five inches in diameter. The gold rings were carried back over the wilderness of mountain and desert to the capital where they were carefully weighed and once again melted down to be poured into molds of several standard sizes. The royal treasury then distributed the ingots or rings to goldsmiths who transformed them into objects of enduring beauty for palaces, temples and tombs.

An ever increasing amount of Egyptian gold was earmarked for the temples and priesthood. At the end of the Old Kingdom, a little more than 4,000 years ago, the priests who administered the thousands of temples dedicated to Egypt's many gods began to accrue wealth and power. The Pharaohs exempted them from taxes and certain other obligations to the state. They no longer had to render a portion of the harvest from temple lands nor provide their slaves and servants as public works laborers.

In the Old Kingdom the Pharaoh had made every decision. But gradually clever priests, whose actions they claimed were always directed by the gods, undermined his supremacy. As Egypt expanded with foreign conquests the priesthoods gained even greater strength. A part of the loot from each military victory was given to the gods as thanksgiving. The priests urged the Pharaoh on to war to gain such booty and when the Egyptians were victorious, the Pharaohs gave them not only booty but tribute gold and land. The most powerful temple, that of Amon at Thebes, once held an estimated 30 per cent of all the land in Egypt. During the reign of Ramses III the temple collected more than 100 pounds of gold yearly as its share of tribute. During the New Kingdom, which began ca. 1500 B.C. and marked the great era of building temples and opulent tombs, the priests of the chief gods were the wealthiest and most influential of Egyptians.

The mortuary temples of former Pharaohs and the large temple complexes were filled with sacred gold. Among the myriad chambers, courts, shrines and columned halls were the saferooms, where treasures were stored and the holy sanctuaries where the gods lived. The great gold statues inlaid with stones and enamel were treated as human kings and queens. The chief priests entered the sumptuous sacred chambers to perform daily rituals which included washing, clothing and feeding the gods. During the Old Kingdom a man's heirs were usually responsible for his funeral arrangements. When people prepared for their own afterlife, it became customary to earmark a portion of one's estate for funeral expenses, including mummification and for payment to special mortuary priests. These priests were responsible for maintaining the vast necropoli, the cities of the dead, and for the perpetual performance of certain rituals. Entombed mummies like the golden statues of the gods required certain services. The wealthiest Egyptians paid the priests for the perpetual performance of daily rituals. Others set aside what they could afford for less frequent rites. In either case, the mortuary priests often embezzled income from the endowments they had been entrusted with and neglected the tombs. In the end almost every tomb was abandoned and sacked.

The very workers who built the tombs and carved out the cliff chambers by day often robbed them at night. Papyrus from the 12th dynasty records the confession of a stonemason named Amenpnufer who broke into tombs at night using copper tools. He told how he and his fellow thieves stripped all gold from the tombs and even tore hundreds of feet of linen bandages from mummies to get at the jewelry and amulets placed between the layers. Sometimes arms and legs came off too. Amenpnufer was caught and imprisoned briefly in the mayor's office near the violated pyramid of Sekhemra Shedtaui. Because looting the tombs was a recognized profession and the thieves were regarded with some sympathy, the stonemason was able to bribe the bureaucrats and return to his double vocation.

Grave robbing has been an industry in Egypt for six thousand years. Until a couple of years ago the maximum penalty for stealing antiquities was an absurd $25 fine. The government, concerned about the destruction of the national heritage, has recently imposed stiff sentences for those caught trafficking in statues, jewelry and other artifacts from ancient sites. It is doubtful, however, that the illegal sales will stop as long as collectors and museums without moral compunctions are willing to pay.

In ancient times as well as today stolen gold found a ready market. A tomb robber's guide entitled "Book of Buried Beads and Precious Treasures" emphasized the high value placed on elaborate stone-set jewelry. Most gold was melted down into anonymous bars or rings and traded on the black market for food, slaves or land. When there was famine or other hardship, gold was bartered for necessities. Some bootleg gold even found its way into foreign markets.

Toward the end of the New Kingdom a period of increasing poverty and lawlessness convulsed the Nile Kingdom as its power and prosperity declined. Thefts from temples and the tombs in the Nile's West Bank necropoli were so common that a royal inquest into royal and private tomb robberies was ordered. Tomb gold was so abundant that it was circulating as illicit currency, adding to the inflation which already plagued the weakened government. Thieves, tortured by the royal inquisitors, spoke of such tempting sights as "the august body of the king wholly covered with gold, his coffins burnished with gold and silver within and without and inlaid with all manner of stones." Most of the accused were craftsmen or minor bureaucrats. Despite punishment by mutilation and banishment to the goldmines of Nubia, the pillaging continued. Finally, ca. 1050 B.C. when almost every royal tomb in the Valley of Luxor had been disturbed, the royal mummies were removed. They were stripped of their jewelry and gold amulets, put into wooden coffins and unceremoniously placed in mass burials at Deir-el-Bahri, where group of them were discovered in 1881.

In 1798 Napoleon embarked on an ill-fated campaign to conquer Egypt. With him were many scholars and an even greater number of adventurers who were attracted to the enduring stone monuments of the ancient Nile Kingdom. Their exploits thrilled Europe. Suddenly there was mania for everything Egyptian. Hairdressers, jewelers, furniture-makers and writers were inspired by the mysterious and exotic land. Those treasures that had escaped previous looters were sought by antiquities hunters, some of whom called themselves archaeologists. They were heedless of the irreparable damage done to the glorious legacy of the Pharaohs. An Italian pseudo-archaeologist who smashed his way into tombs with a battering ram noted, "With every step I took I crushed a mummy in some part or other." His sole objective, like that of many other Europeans and native villagers, was to find the gold and saleable antiquities in such great demand in the drawing rooms of Europe's capitals.

A number of 19th-century archaeologists from Britain, France and the United States sought to unravel the riddles of Egyptian civilization through examination of ancient sites. Scientists who had spent months preparing to enter an excavated chamber usually found they had been beaten to the punch when they looked around to find tomb equipment smashed, stripped of valuable gold sheathing, inlays and panellings of stones, ivory or rare woods. Often they encountered charred pieces of chariots or furniture which had been burned by the robbers. Equally disappointing was finding a tomb with intact artifacts that could add immensely to the slowly growing fund of information about prehistoric Egypt, only to have them all stripped during the night by thieves who had been shadowing the archeologists. The modern looters took everything the insatiable antique merchants would buy—not only treasure but statues, pots, papyri, furniture and weapons, and they callously broke what they didn't value.

After Thutmose I, over 40 more royals had their tombs sculpted out of the tawny limestone cliffs in the Valley of the Kings. Some of them had galleries tunneling hundreds of feet into the rock and a maze of cul-de-sacs and hidden chambers. Royal hopes that the bleak valley in the Western Desert could be easily safeguarded proved false. Gangs of determined outlaws gained access to the forbidden area. They clashed with mortuary guards, sometimes fighting battles among the mummies and their golden treasures.

Only one tomb remained overlooked. It was that of minor pharaoh who was only 18 or 19 years old when he died, possibly murdered by conniving priests and generals. He was interred in a hastily excavated tomb with less splendor than most kings. But when his gold-filled tomb was discovered in 1922, the golden glory of Ancient Egypt was suddenly made manifest. The more than 2,000 artifacts in Tutankhamen's tomb, including the largest single collection of Egyptian goldwork, provided an unprecedented amount of material evidence of ancient Egyptian culture. Tutankhamen, or King Tut as he is called, was born midway between the erection of the Great Pyramid at Gizeh and the Roman occupation, or about halfway through the fruitful 3,000-year span of Egyptian civilization. Not much is known about him. He was the son-in-law of the beautiful Nefertiti and Akhenaten, the revolutionary pharaoh who attempted to make worship of Aten, the sun god, the world's first monotheistic religion. By replacing the multitude of powerful gods with the sole worship of Aten, he expected to restore the traditional supremacy of the Pharaoh which had been badly eroded. The priests whose authority, riches and estates were threatened opposed his reforms. Within a decade he died, and Tutankhamen, about 10 years old, mounted the throne and bowed to pressure to return to traditional religion, although he clung to some aspects of the sun cult. About nine years later, ca. 1352 B.C. he died. His name was obliterated from all records by his more orthodox successors who associated him with Akhenaten. He lay forgotten amid golden treasures until Carnarvon and Carter made one of archaeology's most exciting discoveries and brought him far greater fame than he had ever enjoyed as a Pharaoh.

The first clues of Tutankhamen's existence were found in 1907, when a faience pitcher with his name on it was found under a boulder and several pieces of gold leaf with his name were unearthed from an underground chamber. Because his tomb, so hurriedly prepared, was unusually small and because rubble from the construction of Ramses VI's tomb, dug almost 200 years later, covered the entrance, Tutankhamen's secret was kept. Once traces of his existence were known, however, he became the object of one of archaeology's most single-minded searches. The young Carter and his wealthy patron, Carnarvon, painstakingly worked through 200,000 tons of sifted earth in the Valley of the Kings. Carter dug trench after trench in the area where Tut's pitcher and gold seals had been found, being careful not to work too close to the tourist path leading to Ramses VI's tomb. After six tedious and discouraging years Carter was about to admit defeat. In the off-season for tourists, he could concentrate on digging near their path. Much to everyone's amazement, a flight of stone steps appeared as the workmen shoveled through immense drifts of sand and rock rubble at the base of a high limestone bluff.

Going down the steps Carter came to a sealed door. He didn't open it at once but cabled his patron. When Carnarvon arrived on November 26, 1922, the two men entered. As Carter peered through a small hole made in the doorway, his candle flickered when hot, ancient air escaped. "Presently," he wrote, "my eyes grew accustomed to the light, details of the room emerged slowly from the mist: strange animals, statues and gold—everywhere the glint of gold."

In the four chambers of the small tomb was a dazzling display which mingled the most precious and sacred objects with homely articles, splendid gold jewelry, statues, robes and furniture were piled to the ceiling mingled with fly whisks, seeds, boomerangs, children's games and dried fruits. Most touching of all the artifacts belonging to the young god-king were two tiny coffins containing the bodies of two stillborn babies, presumably his and his wife, Ankhesenamun's.

The tomb had been broken into in antiquity. About a decade after the Pharaoh's death robbers had ransacked the antechamber as Carter discovered from seals bearing the mark of a later pharaoh which soiled the outer door. They took some things including the contents of some 50 cosmetic and unguent vases. They probably poured the precious essences into skins for quick removal, leaving the jars behind. Guards evidently interrupted—the robbers whose footprints were still visible in the 3,200-year-old dust. The guards had not set the room in order but left elaborate gilded chairs, ivory gaming board, weapons, cups and two dismantled golden chariots lying higgledy piggeldy.

The thieves may have taken a number of small treasures, but so much was left that it took archaeologists ten years to catalog and study everything. In the sepulcher's rooms was every pleasant and useful thing a young pharaoh accustomed to the opulent life of court might need or want. Over 400 gilded ushabti figurines waited on him. Military equipment—weapons, chariots, a camp bed and a gold-embellished, folding stool with legs terminating in duck heads—was included, even though he probably never participated in any military campaigns. Golden sentinels and gilded magical animals representing gods watched over Tutankhamen and his burial goods. Almost all the items, whether precious or simple, served both a practical and a magical function.

There were symbols of his royal power such as his marvelous gold throne. There was food—over a hundred baskets of fruit and joints of meat. There was rich clothing, games, bouquets of spring flowers, perfumes and soothing oils, gold ad ivory centerpieces.

Tutankhamen had daggers of gold and one of meteoric iron. He had playthings from his childhood—a slingshot, a painting set and toy chest. There were objects of almost every known material but the warm luster of gold predominated.

Gold covered walls and doors, sheathed life-size statues, formed a massive sarcophagus and was used for pieces large and small worked in a variety of styles and techniques. There was truly an awesome amount of gold. But the sheer volume of precious metal truly meant little by itself. What is significant is the multitude of ways in which it was used. The goldsmiths of Tutankhamen were superb craftsmen. Their sense of design and proportion never failed, whether creating miniature pieces of jewelry or a mummiform coffin. Tutankhamen represented restored order. After the monotheistic experiment with its upsetting changes of the traditional ways, the young king's return to the old religion was comforting to most people.

One area of influence lingered from the so-called Amarna period. During Tutankhamen's reign there had been a revolt against the age-old art forms. Sculptors, painters and goldsmiths were encouraged to break away from the traditional rigid, schematic forms by emphasizing grace, naturalism and a curvilinear style which were new to Egyptian art. Even after Tutankhamen's accession the Amarna art style continued as evidenced in the naturalness of many of the gilded figures carved with elegant realism. Slightly protruding stomachs, flabby chests and active poses in three-dimensional sculptures as well as the unusual attempts at portraiture are elements of the new style.

Tutankhamen's gold throne, one of the most breathtaking treasures, also reflects Akhenaten's influence. The back panel of the chair is a gold canvas where goldsmiths executed a remarkably informal scene of the pharaoh and his childlike wife. The picture wrought in gold, silver, colored glass and costly stones shows Tutankhamen sitting in a relaxed posture on a throne as his queen tenderly applies perfumed oil to his shoulder from a little unguent pot she holds in her other hand. Such an intimate scene had never been portrayed in earlier Egyptian art. The composition is bright with color and richly textured. The royal robes fall gracefully in gold and silver pleats. The flesh is of rosy glass and the ornament details are worked in semi-precious stones of many shades.

The archaeologists found gold amulets, jewelry, bowls, vases and weapons scattered everywhere. Gold-clad chairs, chests, beds, tables and couches shone in the gloom. Highly decorated pectorals, rings, goddesses and golden gods greeted the dazed 20th-century scientists. But the greatest of all the princely treasures in both an artistic and intrinsic sense lay in the burial chamber. In February of 1923 Carter entered the chamber which was flanked by two life-size statues of the king made of gilded wood adorned with stones and dressed in gold sandals. As he opened the door he faced what appeared to be a solid wall of gold.

This was an outer wall of the huge shrine of gilded wood which filled the room. The outermost shrine was almost 17 feet long, almost 11 feet wide and nine feet high. The vast expanse of sheet gold was inlaid with a pattern of blue faience and decorated with magic symbols and text to protect Tutankhamen. In between one heavily gilded shrine and the next was an assortment of funerary objects, including seven magic oars to ferry the king to the Underworld.

Within the second shrine was a third, and within that one, a fourth. Inside the last in this nest of shrines was an imposing sarcophagus carved from a quartzite monolith. It was almost five feet wide and five feet high and nine feet in length. A huge slab of rose granite, which had been broken in two and carefully cemented by ancient craftsman who concealed the seam by tinting it to match the stone, covered the sarcophagus. Two years after Tutankhamen's tomb had been discovered the slab, weighing over a ton, was lifted with a block and tackle. It was a dramatic moment. As the lid was slowly raised, an international cluster of Egyptologists crowded into the tiny crypt. A collective sigh escaped in the stifling atmosphere as they saw nothing more than white linen shrouds. Carter moved the ancient cloth aside revealing an unparalleled sight. "A gasp of wonderment escaped our lips...a golden effigy of the young boy king, of most magnificent workmanship, filled the whole interior of the sarcophagus," wrote Carter recalling the moment when modern eyes first beheld the seven foot long figure.

The gorgeous effigy was actually the lid of a mummiform coffin worked in sheet gold, glass and multi-colored stones, which rested on a low gilded bier in the shape of a lion. Tutankhamen's face and hands were modeled in solid gold of a slightly different color from the thick gold plate of the body. The king's goldsmiths had made his face a true likeness with finely sculpted nose and youthfully vulnerable mouth. The eyes were of argonite and obsidian, the eyebrows inlaid in lapis lazuli. A gold headdress rested on the pharaoh's broad brow, its front embellished with three-dimensional figures of a vulture and a serpent highlighted in red and blue glass. The Pharaoh's arms were crossed on his chest; gold crook in his left hand and a gold flail, both royal symbols, in his right. Two golden goddesses stood guard over the effigy on which lay a small wreath of flowers, still colorful after 32 centuries. The awesome spectacle of the golden sculpture was in a poignant contrast to dried blossoms, which may have been the girl queen's last tribute to her husband.

Since Carter was frustrated by a series of disagreements with Egyptian officials and conflicts in his group, it wasn't until 1925 that he could begin work on the second coffin which lay inside the first. It too was of wood covered with inlaid sheeting of gold. It represented the Pharaoh as the god Osiris, lord of the underworld. The beaten gold of the face is modeled to give a look of sadness and wistfulness, reflecting the young king's untimely death.

Nested inside the second coffin was a third swathed in red cloth and covered with a number of flower wreaths. This innermost coffin proved even more sumptuous than the others. It was not made of heavily gold-plated wood but of solid gold—over six feet, one inches long—weighing 2,448 1/8 pounds, a numbing mass of pure and shining gold. The gold was between 2 ½ and 3 ½ millimeters thick and of a uniform fineness of 22 carats. The face of the third coffin was calm, the eyes tranquil. On the gold headdress were once again the vulture and the serpent, symbols of Nekhbet and Wadjet, the goddesses of Upper and Lower Egypt. The gold body of the pharaoh was tenderly wrapped in the winged embrace of the goddesses Isis and Nephthys.

Sadly, Lord Carnarvon, who had given so generously of his wealth and enthusiasm to the ambitious project, missed the spectacular revelations in the burial chamber. He had been bitten by an insect, probably a mosquito, at the site in March of 1923. The bite on his cheek became infected and he was hospitalized in Cairo. He briefly recovered but then contracted pneumonia and died at the age of 57. At the moment of his death, 2 a.m. on April 5, 1923, Cairo's electricity failed and the city was plunged into darkness. The British High Commissioner, Lord Allenby, ordered an investigation of the power plants. No cause for the simultaneous failure of the generators could be found.

At the same time, Carnarvon's pet dog, who was in London, gave a heart rending howl and died. These weird happenings first fueled the theory of "Tutankhamen's Curse." Subsequent deaths of others connected with the tomb's discovery, including the head of the Louvre's Department of Egyptian antiquities in 1926 caused a sensation. The press pandered to public hunger for stories of the "Mummy's Curse." Other men died. George Jay Gould, the American Midas, died after he fell prey to a mysterious ailment at Luxor. Sir Archibald Douglas Reid, a well known radiologist, signed a contract with the Cairo antiquities department to x-ray Tutankhamen's body and died of an undiagnosed illness almost immediately.

A professor at Leeds University, H.G. Evelyn White, went to Egypt where a priest gave him a set of apocryphal books found in the secret room of an ancient monastery. He took them back to England to study. White killed himself leaving a suicide note which read in part: "I knew there was a curse on me, though I had leave to take those manuscripts...The monks told me the curse would work all the same. Now it has done so." By 1929 almost a dozen people connected with Tutankhamen's tomb and the Valley of the Kings had died under what the press and curse supporters termed "mysterious circumstances." Even the staid *New York Times* stated "It is a deep mystery, which is all too easy to dismiss by skepticism." Claims were rife that Carter and his colleagues had found a table inscribed with a warning that "Whoever disturbs the peace of the Pharaoh, Death will slay with his wings." Carter never found such a clay tablet and there never was an active curse, although the legend enjoyed widespread credence. It troubled Carter greatly. After refusing to comment on it in hopes it would fade away he finally felt compelled to state that it was a "libelous invention." But the rumor persisted and even today is believed by many people.

Carter and his associates, after removing four gilded shrines, two linen shrouds, the massive stone sarcophagus, three more linen shrouds and the three snugly-nested coffins, finally reached the mummified body of Tutankhamen, the first ever found in its original cases. His bandaged head and chest were covered by a stunning portrait mask of beaten gold. The burnished, life-size mask weighed 22 pounds. The expressive eyes were made of obsidian and quartz, the lids and brows accented in lapis. Blue glass alternated with gold to make up the stripes in the traditional headdress. The customary vulture and serpent were of solid gold, embellished with cornelian, lapis and blue faience. The cobra's eyes were formed of gold cloisonné inlaid with translucent quartz over red pigment. On his chest was a broad collar, the ends of which terminated in gold falcon's heads. The collar was encrusted with quartz, green feldspar and lapis lazuli arranged with an unerring eye for the most striking design.

The Pharaoh's body lay under the mask, wrapped in hundreds of feet of linen. Tucked into the bandages were 143 gold and jeweled charms and amulets to protect him against evil forces. Only his hands and feet were uncovered, each finger and toe sheathed in a gold cover. On his feet were golden sandals. On his shaved skull he wore a linen skullcap with a delicate design picked out in miniature gold and faience beads. Over it was a linen headdress held in place by gold bands and a lovely gold diadem encircling his brow.

Around his neck was a jeweled gold necklace and on his breast Carter found a series of small pectorals of the finest cloisonné work, arranged in 16 layers. There was more. On the Pharaoh's hands were two gold rings, another 13 were tucked in the linen folds. All of them were heavy, handsome rings with cartouches, the characteristic oblong figures bearing the names of the god-king. Eleven splendid bejeweled bracelets had been placed on the mummy's arms with two girdles of gold and jewels about his waist. A gold-handled dagger with an unusual blade of iron was at his hip and an apron of inlaid gold covered his legs.

As if this staggering amount of gold weren't enough, Carter discovered even more as he explored a small room reached through an opening in the west wall of the burial chamber. He had exercised his usual self-discipline by sealing off the room at the time the royal mummy was discovered. Two years passed before he turned his attention to what has been dubbed the Treasury. This tiny room contained hundreds of ushabti figures, a fleet of model boats, the coffins of the two infants and a large gilded wood shrine in the center. Scenes of religious significance with accompanying text were carved in the surface, which was heavily gilded. Four golden goddesses, facing inward and with arms outstretched, guarded the shrine's contents. Within, Carter found the mummified internal organs of Tutankhamen which had been removed, cured in a solution of natron and aromatic spices and then bandaged. Each organ was put in a miniature coffin of beaten gold, inlaid with carnelian and colored glass, then in a canopic jar of alabaster in the shape of the pharaoh's head. The slender goddesses at the sides of the shrine were guardians of the respective organs: Neith watched over the stomach, Selkis the intestines, Nephthys the lungs and Isis the liver.

Scholars throughout the world were delirious with joy at the wealth of information Tutankhamen's grave goods provided about

Egyptian culture. The public was also wildly excited but chiefly at the sheer volume of golden treasure massed together in the small rock-chambered tomb. People everywhere marveled at the variety, beauty and consummate craftsmanship of the thousands of artifacts, which reflected the sophisticated world of one of history's crowning civilizations.

Tutankhamen's tomb furnished a unique glimpse into the opulent 18th Dynasty, which was marked on the domestic front by the construction of imposing temples and lavishly-filled tombs. Less spectacular but very compelling evidence of Egypt's activities in the expansionist 18th Dynasty comes from a series of eloquent bas-reliefs in one of the most beautifully designed of all Egyptian temples.

Carved in the walls of the colonnaded mortuary temple built at Deil el Bahri for Queen Hatshepsut are scenes depicting the major achievements of that most remarkable woman's reign. Hatshepsut, the strong-willed female pharaoh, ruled for 20 years and considered the expedition she sent to the distant gold lands of Punt one of her greatest accomplishments. She was the daughter of the empire builder Thutmose I and such a brilliant, clever woman that she managed, against great odds, to seize power from her stepson whose regent she was, and maintain peace for two decades while expanding international trade and planning ambitious building projects.

There had been a couple of other women rulers in Egypt but none who had dared assume the sacred kingship and proclaim herself divine. The ambitious Hatshepsut mounted the throne in 1503 B.C. and had the nerve to adopt male ceremonial attire. In some statues she is even depicted with the traditional false beard of the pharaohs. A woman of great courage and ability, she assumed the titles due the divine ruler of the world's greatest empire. "Son of the Sun, Mistress of the World, Golden Horus," she called herself as well as by 80 more honorifics.

The Egyptians had been an aloof and insular people in their early history, content to live on the lush borders of the Nile, surrounded by a sea of sand on three sides and the Mediterranean to the north. This isolation aided in the formation of a closed society with rigid, conservative overtones. The desert sea was a buffer against both invading peoples and the introduction of revolutionary foreign ideas and customs.

Gradually, however, the Egyptians had sought out those things they came to consider essential—primarily wood, building stone, copper and gold. Foreign involvement was limited to trade in the Old Kingdom when the first recorded expeditions voyaged to mysterious Punt in the mid-3rd millennium B.C. Today the precise location of Punt is a matter of some debate. Some think it may have been the area of Cape Guardafui on the Somali coast where frankincense and myrrh grew in abundance. More likely, it was further down Africa's east coast in the auriferous area between the Zambezi and Limpopo rivers where the enigmatic ruins of Zimbabwe lie among ancient gold mines littered with human bones. At one time the majestic stone monuments in the Rhodesian jungle were attributed to ancient builders from abroad. But it now seems certain they were constructed by the once powerful Rowzi tribe in the period between 1100 and 1600 A.D.

In Egyptian texts and inscriptions Punt is described as "paradise on earth," source of much gold and spice. Under Ramses III, who reigned in the 12th century B.C., about 300 years after Hatshepsut, gold-seeking ventures to Punt were well organized on a large scale. Thousands of merchants and seamen traveled there and back, suggesting the route was well known. About 2300 B.C. a helmsman named Knemhotep died and was entombed near the First Cataract of the Nile. He had been a man of means whose wealth, according to tomb inscriptions, came from Punt. He was in partnership with a captain named Khui, and together they had brought back gold, ivory, ebony and incense.

Hatshepsut deserves the acclaim she accorded herself for reestablishing and enlarging international trade connections, which had withered during the humiliating, instable century or so when Egypt had been ruled by foreign invaders from Asia. The Hyksos, fierce nomadic tribesmen, had beer driven from Egypt about 80 years before Hatshepsut became Pharaoh. They had been cruel masters but benefited Egypt by breaking her former culture isolation, thus paving the way for a renaissance that blossomed in the new kingdom.

Beginning with the 3rd millennium B.C., Egypt's lust for gold had spurred one prospecting expedition after another as the alluvial deposits, where gold nuggets had once been found in such profusion, were exhausted. Mines in the deserts were yielding reduced amounts by Hatshepsut's time. Nevertheless, by most standards there was still a great deal of gold to be found in Egypt's mines and placers. But the world's richest queen needed infinite mounts of gold for her grandiose schemes. She planned an offering to the great god Amon-Ra. It was to be the greatest offering ever made, something worthy of both the god and the divine queen who was earth's mightiest ruler. She confided in her chancellor and favorite, Senmut, a plan to erect two towering gold obelisks, which would flash their light to heaven over the high walls of the gigantic Karnak temple complex at Thebes.

Senmut, who was a practical man, curbed his sovereign's extravagant dream. He prevailed upon her to abandon the idea of pillars cast in pure gold. Instead two colossal shafts of granite, each nearly hundred feet high, were made. The obelisk had been associated with the rays of the sun god long before Hatshepsut. The ancient sun worshippers believed that they were conductors, transmitting the sun's power into the earth so that it would be fruitful. Hatshepsut's granite monuments weighed 700 tons apiece and were transported from the quarry at Aswan to Karnak on a mammoth ceremonial barge covered with gold and pulled by 30 boats powered by 900 oarsmen.

Hatshepsut wanted to sheathe the pillars with electrum, the pale gold with which she dusted her face. But Thuty, the Overseer of Gold and Silver, protested that even that was too extravagant, so she settled for covering the pinnacles of the twin monuments with heavy gold plate. Reliefs in Thuty's tomb show him dipping into the royal treasury to give his queen the 12 bushels of gold to cap the obelisks. Her own inscriptions claim she measured the gold herself. "Hear ye!" she declared, "I gave for them the finest of electrum which I had measured by the bushel like sacks of grain." She exulted that the glow of the obelisks could be seen "on both sides o± the river...Their height pierces to heaven... Their rays flood the two lands when the sun rises between them. Never was done the like since the beginning."

Her proud words are still visible carved into the stone of the one remaining obelisk which was partially defaced. After her death, her stepson, Thutmose III who succeeded to the throne, made every effort to obliterate the memory of the great female ruler. One obelisk crumbled. Fragments of it were carried away for use as millstones turned by blindfolded oxen along the Nile. Other pieces lie where they fell. The remaining obelisk still towers skyward, the tallest ancient construction in Egypt and an imposing reminder of one of history's most outstanding women. The gold was stripped years ago but Hatshepsut's boast evokes the vanished spectacle of her twin obelisks. "You who after long years shall see these will say, 'We do not know how they can have made whole mountains of gold.'"

Hatshepsut needed gold for the Karnak monument and also for the financing and decoration of her mortuary temple at Deir el Bahri. She assigned Senmut to design and supervise the building of the memorial. He was an ingenious architect and executed a novel design built against the base of a towering cliff on three levels. His plan included landscaped gardens to be filled with exotic trees and flowering plants and a series of graceful porticoed terraces. Hatshepsut's story was told by almost 200 bas reliefs and statues. They were designed as propaganda to strengthen her claim to the traditionally male kingship. In them she tells of her divine and royal ancestry, of the obelisks and of the expedition she dispatched to Punt.

She was motivated by the greed for gold and other luxuries as well as by the desire to assert her authority and show her link with the great Pharaohs of the past who had traded with the fabulous far-off country. The queen sent a fleet of five big ships under the command of the versatile Senmut. Reliefs show the arrival of the ships at Punt when the Egyptians presented the natives with many gifts. The Puntite chiefs are depicted receiving tools, weapons, bangles and bracelets, which were probably of colorful glass paste and some semi-precious stones. The Egyptians were unlikely to have sent gold to a country which reportedly brimmed with it. At a great communal banquet set up in the open and presided over by an enormous statue of Hatshepsut, the Puntites were plied with Egyptian beer, vegetables and wine—"all good things of Egypt which the queen has commanded."

In return the chiefs of Punt bestowed an amazingly generous bounty of African luxuries on their foreign visitors. The reliefs at Deir el Bahri illustrate porters bent double under load after load as they fill the Egyptian ships. The great vessels are seen riding low in the water under the weight of "the costly marvels of the land of Punt; all goodly fragrant woods of God's land, with fresh myrrh trees, with quantities of ebony and ivory, with the green gold of Emu, sweet smelling resin, eye cosmetic, with apes, monkeys, greyhounds; furthermore, with skins of the southern panther and natives of the country with their children." Hatshepsut concludes that "never was brought the like of this for any king who has been since the beginning."

It took two years for the vessels which had sailed down the Red Sea into the Indian Ocean to return with their precious merchandise. The female Pharaoh was obviously delighted with the eagerly awaited cargo. "The people of Thebes rejoice because of the greatness of the marvels," wrote her scribe. "Punt has been transferred to Thebes."

The spirited reliefs show the royal treasurer presiding over the weighing and dividing of the African gold, which arrived in the form of dust, nuggets and gold rings such as those shown worn up to the thigh by one of the princes of Punt. With a gold scoop "her majesty acting with her two hands" assists as four men remove gold dust and nuggets from treasure chests. The gold-loving priests of Amon Ra stand nearby waiting for their share. The principal figure is Hatshepsut. Anointed with fragrant oils "she exhaled the odors of the divine dew, her fragrance reached as far as Punt, it mingled with the odors of Punt." Powdered gold dusted her skin which was "like kneaded gold, and her face shone like stars in a ceremonial hall."

Hatshepsut's deeds live on in spite of her successor's efforts to deny them. But 35 centuries later there are none of the golden treasures she so loved—none of the green gold of Emu, which she must have parceled out to goldsmiths to make into ornaments and exquisite articles for palace, temple and tomb survived the melting pot.

Egypt had another extraordinary queen some 1,400 years later who was as ambitious as Hatshepsut and as enamored of gold. She was the seductive young Cleopatra, descended from Ptolemy the Macedonian who became king of Egypt after Alexander the Great's death. Years of Persian tyranny followed by Macedonian rule had weakened the Nile Kingdom. Cleopatra sought to bolster her beloved country's sagging fortunes, first by snaring one powerful Roman and then another. When she realized that Caesar had been stabbed to death in Rome, she welcomed Mark Antony's attentions as a way to preserve the integrity of her kingdom. But in 31 B.C. Antony's powerful rival, Octavian, led the Roman fleet against the Egyptians in the battle of Actium. The Romans prevailed and Antony and Cleopatra fled back to Egypt. In despair the proud queen held a poisonous asp to her breast. She preferred death to the humiliation of being paraded in chains, amid displays of all her gold, through the streets of Rome. With her suicide the enduring golden civilization, which consumed as much as 70 per cent of antiquity's gold production, perished. In 30 B.C. Egypt, the eastern paradise, became the crowning jewel in the expanding, gold-greedy Roman Empire.

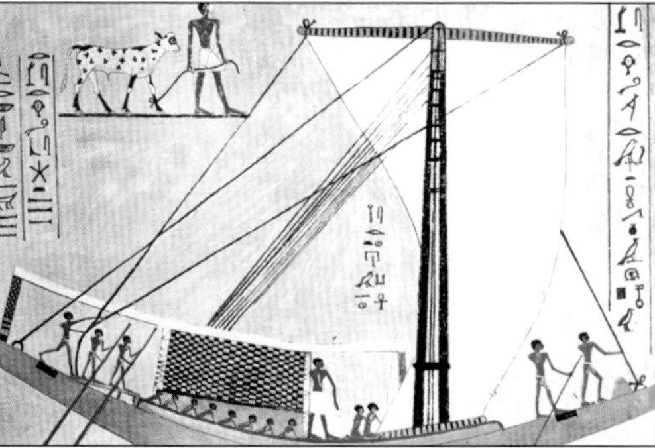

(Above, left) Egyptian wall painting of goldsmiths pouring molten gold into prepared molds with others hold tongs and blowpipes used as bellows. *(Above, right)* Egyptian ship of the 6th Dynasty (2345 - 2183 B.C.).

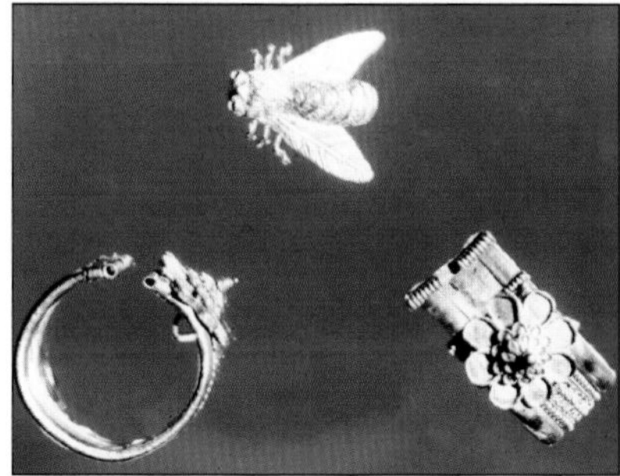

(Above) Egyptian gold jewelry ca. 2200 B.C.

(Right) Pharaoh Tutankhamen's (King Tut) golden canopic coffin detail.

(Below) The interior of Pharaoh Tutankhamen's tomb.

(Above, left) Gold-gilded shrine found in the antechamber of Tutankhamen's tomb.

(Right) Tutankhamen's Throne with footrest—1323 B.C.

(Above) Gold and precious stone pectoral from the tomb of pharaoh Tutankhamen, ca. 1350 B.C.

Buckle of open goldwork depicting Tutankhamen on a chariot, discovered in his tomb.

The back panel of Tutankhamen's gold throne is a gold canvas where goldsmiths executed a remarkably informal scene of the pharaoh and his childish wife. The picture wrought in gold, silver, colored glass and costly stones shows Tutankhamen sitting in a relaxed posture on a throne as his queen applies perfumed oil to his shoulder from a little unguent pot.

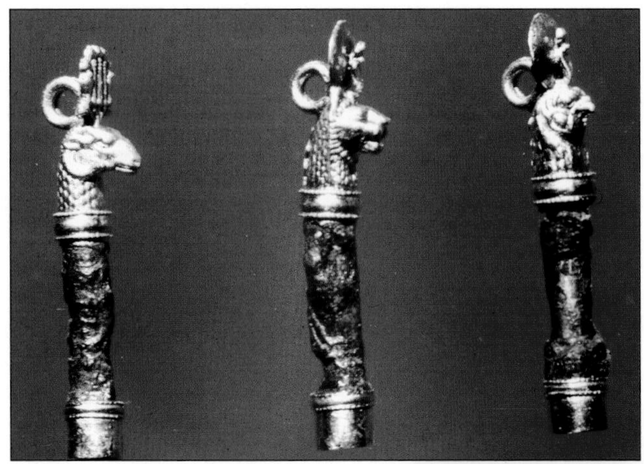

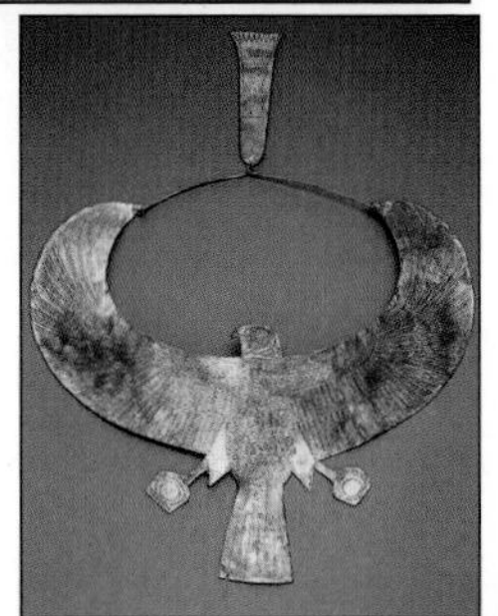

(Above) Amulets of Egyptian gods ca. 400 B.C.

(Right) This falcon collar of gold is from Egyptian Pharaoh Tutankhamen's mummy.

(Below) This Persian gold bowl belonged to King Xerxes, 5th century B.C.

Chapter 5

The Gold Of Troy

Ancient Egypt had no peer as the paramount producer, importer and consumer of the sun's dazzling metal. But she did have rivals who also valued gold as the supreme attribute of power and wealth. While slave armies hacked through the quartz reefs of the Egyptian gold mines and the insatiable Pharaohs mounted expeditions to bring back still more gold from faraway lands, the accumulation and use of gold was spreading to other lands.

The Second Millennium before the birth of Christ was a time of growth around the Mediterranean. Regional Bronze Age chieftains consolidated power. The development of metal industries stimulated trade and communications. As their kingdoms grew they joined in the trade which had metals—gold, copper and tin—as its primary objective. Island kingdoms and mainland principalities that enjoyed the good fortune of proximity to deposits of the coveted metals or that controlled strategic trade routes launched themselves into international commerce, trading with or for gold.

The Minoan civilization of Crete was the first high civilization of Europe and the first great gold culture outside the East, but until a century ago the modern world had never heard of it. The first evidence to historians of the dazzling civilization which flourished on Bronze Age Crete came from Egyptian records. A fresco found in the tomb of an Egyptian high official at Thebes shows slender, dark men in foreign clothing who are bearing gifts for the Pharaoh. On their shoulders are bronze weapons, ingots in the shape of ox-hides and one man carries a gold bull's head. Its inscription calls the men "Keftiu" from the "Isles of the Great Sea."

In the ruined city of Akhenaton, briefly Egypt's capital under Ikhnaton's regime of sun worship, archaeologists found another reference to the "Keftiu." In the House of Rolls they came upon 300 clay tablets inscribed in cuneiform characters. They were letters and documents of the 14th and early 13th-century B.C. addressed to Ikhnaton and his father by rulers of several foreign states. The famous Tell el Amarna letters, several of which are requests for eastern gold, coincide with the epoch when Egypt's gold production was at its peak. The pharaohs were exporting a great deal of gold, not only from their own mines but also gold which they had imported from Arabia and the East Coast of Africa. One of the best customers for the Egyptian gold was the mysterious "Keftiu."

About the middle of the Second Millennium Cretan maritime supremacy was eclipsed by the Mycenaeans who took control of the south Aegean and the island settlements formerly under Cretan domination. The Minoan civilization was in the process of disintegration but an Egyptian list of tribute which the Minoans rendered to the pharaoh shows that the Aegean island was still wealthy and its artistic level high. Among the treasures presented to the Egyptian king were "bars of gold and silver, silver rings, baskets filled with lapis lazuli, ornamental vessels, some shaped like heads of bulls, lions, dogs, griffons and goats made of gold, silver, copper, and silver-gilt chains of red and blue pearls, and daggers."

Archaeologists and scholars were puzzled by the origins of the "Keftiu." Who were these slim men who had such an abundance of treasure? The answer was revealed by a near-sighted English scholar who was the keeper of the Ashmolean Museum at Oxford. Sir Arthur Evans went to Crete at the end of the 19th century to track down the origin of some engraved seals with hieroglyphic symbols in a completely unknown script which he had seen at Athens. He then decided to excavate and made a find that revolutionized history. On the second day of excavating he discovered the Bronze Age palace at Knossos—the palace of King Minos—whose name Evans gave to the brilliant civilization which flowered on the fair isle of Crete.

His excavations revealed the most elaborate prehistoric building in the world—a palace complex of immense size, more than 1,200 rooms, a veritable maze of courts, reception halls, pillared galleries, bathrooms, ceremonial chambers, domestic quarters, warehouses and workshops. The plastered palace walls were covered with hundreds of marvelous frescos of Minoan life. Among the joyous scenes were agile young women and men shown leaping gracefully over the horns of charging bulls in ritual games. This was the first bit of evidence linking the Greek legend of Minos and the terrible Minotaur who devoured human victims in the subterranean labyrinth of the royal palace to historical fact.

The Classical Greeks had many legends that have been passed down through the generations, cherished as the fanciful myths of the fathers of our own Western Civilization. Very few believed there was more truth than imagination in the legends of ancient Greece. Only in the past century has the archaeologist's shovel revealed the nucleus of truth from which such legends sprung. Evans on Crete and Schliemann at Troy and Mycenae brought the gold-rich, multi-faceted Minoans, Trojans and Mycenaeans out of the mystic fog into the sun-drenched light of Mediterranean history.

One day long ago, goes one of the beloved Greek tales, a group of lovely girls were playing in a flowering field at the sea shore. By far the most beautiful of them was Europa who was the daughter of the powerful king of Sidon on the Syrian coast. The breeze rippled her long silky hair as she gathered flowers in an exquisitely engraved basket of gold which had been made by Haephestos, the goldsmith of the gods. The great Zeus, seated up on Olympus, looked down on her. He had a way of falling in love easily and was immediately stricken with love for the princess. Zeus was very clever at obtaining the object of his desire. In this case, he transformed himself into a great bull, which suddenly appeared before the princess in the meadow. He lay at her feet and gazed at her with great mournful eyes. Europa was completely won over by the gentleness of such a magnificent beast. She ignored her friends' warnings and climbed on his glossy back. He jumped up, plunged into the waves and swam with her to Crete, his own special island. She lived there happily with him and bore him splendid sons.

The legend of Europa to which some people have listened for some 2,500 years is, of course, the explanation of how civilization reached Europe from the eastern and southern shores of the Mediterranean. And Evans proved that Crete was indeed the first European high civilization.

One of Europa's sons was Minos who became the great ruler of Crete and is credited in legend with establishing the world's first armed navy. His island kingdom was impregnable; his mighty fleet of warships and merchant vessels was invincible. Crete was the stepping stone to all the eastern Mediterranean. His naval power was such that Crete controlled the seas and sailed far to trade for gold while warships kept invaders at bay. He established his magnificent palace high above the sparkling deep blue Aegean and left it unprotected by fortifications or encircling walls.

According to ancient legend Minos' only son had been treacherously murdered by the jealous king of Athens. Minos conquered Athens and vowed to raze the city and kill every one of its inhabitants unless the city agreed to his tribute demand. The Cretan king stipulated that every ninth year seven of Athens's best young warriors and seven of the loveliest maidens be shipped to the palace at Knossos to be sacrificed to the horrible Minotaur. The Minotaur was, as everyone knows, a monster—half man and half bull. He was the offspring of Minos' wife and a sacred bull which Poseidon, the sea god, had presented to Minos. Minos ordered the famous architect and inventor Daedalus to build an underground "palace of confinement" from which there was no escape. The massive palace above trembled with the roars of the crazed monster and the anguished cries of his young victims. Finally the Athenian hero, Theseus, slew the Minotaur with the aid of Minos' daughter the tender-hearted Ariadne who gave him a thread to follow out of the labyrinth.

So much for the myth, dismissed as the imaginings of an inspired storyteller. Evans, an infinitely patient man, dedicated thirty years of his life and his large fortune to filling in the outlines of the bold, vivacious maritime people who practiced the art of graceful living long before any other Europeans. He proved the existence of a powerful Minoan dynasty and a cult of bull worshipers.

Homer's Odysseus called Crete, "hospitable, handsome and fertile." The bold, bright Minoans evolved from the blending of the Stone Age island natives and immigrants who sailed from Anatolia about 3000 B.C. The farmers and herders who settled on Crete brought their culture with them. They were cave dwellers who cultivated grapes, corn and olives with a powerful mother goddess as their chief deity. For centuries these peaceful island people at the crossroads of the Aegean were able to concentrate on developing their agricultural practices. Their animals multiplied and there was a trade surplus of wine, grain and especially olive oil. The Cretans, having descended from pastoral ancestors, were not naturally sea- oriented. They depended on Egyptian ships or Egyptian captains flying Minoan colors to carry their exports and bring back the raw metals they craved. The island had, aside from a negligible amount of gold, almost no native metals. But they imported lots of it, and Minoan goldsmiths and armourers became famous.

They worked gold, copper, tin and silver into highly prized manufactured products traded throughout what the Greeks called the Oikoumene, the part of the world considered habitable. No other nation could surpass the charm and delicacy of Minoan gold work or the quality of the bronze weapons and tools. They were the first people to grow rich in the arms industry. In addition to the products of their metal industries the Minoans exported fine pottery and agricultural products, especially olive oil, which was used in great amounts for religious rituals as well as for cooking and cosmetics. In return they gained raw metals for their famous bronzes, gold, and silver, Syrian dyes, eastern incense, golden amber, which Greek legend identifies with the tears of the sun's daughters, rare woods, ivory and semi-precious stones.

About 1800 B.C. a wild horde of Asiatic nomads called the Hyksos overran Syria and Palestine. They conquered part of Egypt and for at least a century established barbarian rule in the Nile kingdom. During this time the Minoans developed navigation and trading. Their merchantmen traded throughout the Aegean, establishing commercial settlements on a number of islands including Rhodes and Cyprus. The Minoans traded with the mainland Greeks, particularly the people who called themselves Aechaeans. These included those whom Homer called "Golden Mycenae" whose culture and goldwork were deeply influenced by the Minoans who they eventually routed as masters of the sea. For three and one-half centuries the resourceful Minoans were unchallenged. They traded with parts of Asia Minor and the Levant. They foreshadowed the daring exploits of the merchant sea-kings of Phoenicia sailing as far as Tartessus in southern Spain for gold and silver. They may even have ranged as far north as Britain's southern coast where Mycenaean goldwork and weapons have been found in the graves of the chieftains who profited as middlemen in the trade for Irish gold and Cornish tin. Red semiprecious stones from Cornwall have been found in jewelry excavated at Knossos. Perhaps they were brought back by a Minoan captain for his lady.

Digs in locations such as Falmouth in Cornwall, Romania and southern Spain have turned up Minoan gold jewelry and metal ox hides in gold, silver, copper and bronze stamped with the sign of the Cretan mint. They were accepted everywhere as units of value because Minos' might was already assuming mythic proportions.

The primary source of Cretan gold was Egypt. Throughout the Second Millennium the world stock of gold was growing. More and more of it was being attracted to the Near East and the Mediterranean. As Egyptian might faltered or the Pharaoh struggled with internal problems the tremendous supplies of gold, which had been increasing for several thousand years, were diverted to prosperous people who wanted it but had none of their own. The Minoans, in time, supplemented Egyptian gold with gold from other sources. Spain and Ireland supplied some, but far more came from the gold-rich area of southwest Europe and several Aegean islands. Miner's picks made of copper have been found by the dozens in prehistoric shallow gold diggings in the Balkans and Carpathians indicating that European gold mining was at least somewhat organized in the Second Millennium.

Gold deposits existed in an area stretching north from the Aegean island of Thasos through Thrace and Macedonia (the gold which made Philip of Macedon, Alexander the Great's father, enormously wealthy and provided the material to coin Europe's first gold currency). More gold was found in the Balkan area, across the Danube through the Carpathians to the east and westward to the area where the Alps terminate at the head of the Adriatic. Cretan ships probably sailed fairly regularly to the northernmost Aegean ports to trade for the Balkan and Carpathian gold collected there.

Not far from Crete there was gold on the island of Cyprus, which came under Cretan domination and was the chief supplier of copper. Cypriote gold deposits weren't exhausted even by the 15th century A.D. when the island came under control of the Venetian Empire. The island of Siphnos also had gold and silver, so much of it in fact that it was a source of precious metals during classical times.

The Minoans enjoyed what must have been the pleasantest culture of prehistory and one we can easily relate to. Archaeological findings have shown what happy and cultivated lives these clever islanders led. Their life and art celebrated the familiar world of sun, sea and fertile nature. The Minoan culture, which evolved from native, Anatolian, Egyptian and Syrian influence, was the polar opposite of Mesopotamian culture with its emphasis on fear of the omnipotent gods and terrible bloodlust. The Minoan world was a place where man felt at home, the center of attention. In contrast to Egypt or Mesopotamian civilization there was no great emphasis on death and the afterlife. More than any other prehistoric culture Minoan Crete celebrated human life with its simple pleasures. In Egypt the architecture was designed on a scale to humble the commoner and impress even the gods. In Mesopotamia towering ziggurats were raised to honor the gods and keep them from wreaking vengeance on helpless men. On Crete, architecture's focus was on comfortable living. Although the royal palaces of Knossos, Mallia, Phaestos, Gournia and Zakro were elaborate and magnificent structures, luxury was not restricted to aristocrats. Merchants and wealthy farmers lived in comfortable but not ostentatious houses, constructed with timber frameworks to protect against the shocks of earthquakes. Brightly colored paintings enlivened the walls of houses, some equipped with sewers and drains.

The houses were well-appointed and their inhabitants dressed with elegance and indulged in a taste for light hearted opulence. During the peak of Minoan hegemony Minoan culture and artistic life centered on the palace complexes. Large portions of the palaces were designed as administrative offices where clerks and scribes kept track of commercial ventures on clay tablets in the curious Minoan script, which had first lured Sir Arthur Evans to Crete. There were large storerooms where gold work, bronzes, pottery and agricultural products were kept. One such warehouse had 18,000 clay amphorae in which olive oil was stored. Another had pipes through heated water in cold weather circulating evidently to keep the oil from congealing.

Goldsmiths, armourers and potters had workshops and quarters within the complex. Goldsmiths were held in awe and treated with deference. They worked with skill and were inspired by the radiant world of generous land and bountiful sea. Egyptian art like Egyptian life was static, fixed for eternity. Very early the Nile goldsmiths developed stylized representations of natural forms and stuck with them through the centuries. Minoan art is far more passionate, infused with a love of life that leaps and shimmers like the porpoise and flying fish on golden cups and colorful murals. The forms and rhythms of nature moved the sensitive Minoan goldsmiths to celebrations of creation which charmed even the death-oriented peoples whose own smiths were far more solemn.

The art of the metal worker was surrounded by an aura of magic. The smiths in their palace workshops made no effort to enlighten the uninitiated. The mystery gave rise to legends which the later Greeks preserved. Goldsmiths' skills came from supernatural beings such as Dactyls who were said to live on Crete's Mount Ida. They came to the aid of bronze workers and taught men to use iron. Creatures called the Curetes and Corybantes taught armourers their secrets, and the magical Cabeiri protected all metalworkers. According to tradition the first statues were made by the Telchines, the magical beings in charge of working gold and silver. By the time of Homer a goldsmith had been admitted into the company of the gods. The god Hephaestus, known as Vulcan to the Romans, was the only ugly immortal. He was also lame and created female robots out of pure gold to serve as his assistants. From his workshop came all of the gods' weapons, appointments, and jewelry. He made gold sandals for Mercury, golden apples, golden balls, whatever was desired. The Athenians honored him highly. Hephaestus was patron of the crafts along with Athena, and he was the presiding deity at the important ceremony at which a young man was admitted to the city organizations. He was also the god of fire and volcanic eruptions were often attributed to his workings at an underground forge.

Homer described how Hephaestus made armor for Achilles and decorated the polychrome shield with: "a vineyard teeming plenteously with clusters, well wrought in gold; black were the grapes, but the vines hung all on silver poles. Around it ran a ditch of blue cyanus, and round that a fence of tin. Also he wrought a herd of kine, fashioned of gold and tin. And herdsmen of gold were following the kine, four of them, and nine swift dogs came after. But two terrible lions had seized a loud-roaring bull that bellowed loudly as they seized him, rending the great bull's hide and devouring his vitals and his dark blood."

The smith's greatest fame came from creations in gold and ivory, the sculpted pieces of ivory set with gold called chryselephantine which the Minoans originated and the Classical Greeks regarded as the summit of artistic achievement. The famous sculptors of ancient Athens, Pheidias and Polycleitus, whom we know as-workers in marble, were best known by their contemporaries for chryselephantine objects, which sadly have not survived. A few charming pieces of Minoan gold and ivory exist. They show how strikingly the vibrant Minoan art differed from the more stereotyped styles of their contemporaries. There are three statues in particular, a girl bull-leaper, a gold and ivory boy and a lovely snake-handling goddess figurine, which are beautifully modeled. Gold is used sparingly but to great effect, forming elements of dress and jewelry and forming the nipples of the girl bull-leaper and the "snake goddess" in her characteristic Minoan breast-baring costume. Minoan women were important members of society, co-equal with men and active in games, government and the crafts. Men and women shared jewelry styles and no doubt had great influence on goldsmiths' designs.

The earliest Minoan gold work found in tombs was simple. It bore the imprint of the Anatolian homeland where goldsmithing was colored by Mesopotamian influence. Sheet gold was cut in the forms of flowers, leaves, beasts, birds and sea creatures. During the Middle Minoan period starting around 2000 B.C., goldwork became much more elaborate. The goldsmiths of Crete were technical geniuses. They practiced the full range of techniques working with lacy filigree, sheet gold, repoussé, hammering vessels in the round, soldering, creating shimmering patterns of golden dew with tiny golden drops affixed to a golden surface. They loved inlays. Daggers were damascened by inlaying various metals such as gold alloys of different hues, silver and black niello, and a mixture of mead, silver, sulfur and copper, to create elaborate scenes of hunting or warfare.

Vessels carved out of steatite, a kind of soapstone, were covered with the thinnest of gold foil which adhered to the curving decorations of the stone surface. Gold foil was often inlaid with amber, crystal and semi-precious stones. The Cretan smiths learned how to gild bronze and how to plate cups and bowls of silver by bonding the two metals. A royal tomb at Dendra contained a cup made in the 16th century B.C. that has an interior of smooth gold. Galloping bulls in high relief chase around the exterior side of silver. The Dendra cup is a marvelous piece in which a celator with deep feeling for marine life has worked a repoussé scene with dolphins and four great octopi lying among rocks and undulating seaweed. Chasing and embossing were the forte of the Minoans and there are hundreds of examples of these techniques. A chased bowl was made by first cutting a design into a matrix of bronze or stone such as granite or basalt using chisels, borers and punches. A sheet of gold foil was then fastened over the matrix with a leather tong, and a punch of wood or horn employed to hammer into all the crevices of the carved design. The more refined pieces of celature were then chased on the surface after they had been embossed.

The Minoans were spared the violent upheavals that affected so many developing cultures in the Bronze Age. The naval prowess of the Cretan fleet protected the island for centuries and fortune smiled on the Minoans. It is easy to imagine the lovely ladies pictured on so many pots and walls as they try on the golden ornaments they have ordered from the island goldsmiths. The provocative costume of bared bodice and colorful flounced skirts covered by an apron lent itself to the display of jewelry. Minoan women fixed their long hair in elaborate coiffures which were set off by long golden earrings. They wore earrings, necklaces, pendants and rings with intaglio seals. Gold was transformed into bulls' heads, deer, shells, flowers, bees and countless other natural forms. One circular pendant is formed of two dogs on the backs of a pair of monkeys. Another has two human heads in repoussé, which have long swirling hair curling about the miniature ears. Some are in the form of circles ending in animal heads; inside one of these, filling the center are two lions devouring sheep. Other jewelry of the period includes necklaces with pendants representing alternating bulls' heads and crouching lion cubs, hair ornaments, and little gold plaques in various shapes pierced with tiny holes for attachment to the skirts of the elegant ladies.

The Minoans were capable of performing seemingly impossible feats of working gold on a miniature scale. Evans found a minute lion, duck, fish and a heart in the Royal Treasury at Knossos. Even under a microscope they appear perfect marvels of gold-soldering and granulation. One of the golden artifacts from Knossos is especially beautiful because it combines beauty of workmanship with the characteristic Minoan feeling for rhythm and harmony. It is a royal sword with the pommel in the form of a golden acrobat who arches his body back to effortlessly touch his toes to his head. Evans also found a magnificent game board in the palace that was made of gold, silver, lapis—lazuli, rock crystal and ivory inlays.

A unique form of Minoan goldwork was the golden double ax or "labrys." The religion of the Minoans is rather obscure, but it seems the Cretans worshipped a great mother-goddess who took many forms and was worshipped not in temples but in sanctuaries in caves and on hill tops. Archaeologists have discovered votive deposits in these shrines and caves where figurines, double axes and other objects have been found along with a religious symbol called the horns of consecration, which represent the horns of the bull. The "labrys" from which comes the word labyrinth was the most sacred symbol of the Minoan religion and countless votive axes made of gold, silver, ivory, bronze, marble and copper were made as votive objects. So many ritual axes were found in the shrines on the coast east of Knossos that some archaeologists theorize that they were once exported to the mainland as part of a drive to convert others to the mother-goddess.

Minoan ideas were exported along with material culture and inspired other Aegeans. The chief heirs of the spirited Minoans were the Mycenaeans of continental Greece. They were a fierce people renowned for their excessive love of gold and battle. Grown rich and powerful through piracy and warfare as well as legitimate trade, the Mycenaeans gained control of Crete in the mid 15th century B.C. They overwhelmed the Minoans following the devastation caused by a great volcanic eruption on a neighboring island. For two and a half centuries the Mycenaeans and the surviving Minoans formed a hybrid civilization among the shattered, fire-blackened ruins of the once proud palace towns. Then the fading light of the first European civilization was extinguished forever in a mysterious cataclysm caused by native revolt, invasion from the mainland or another natural disaster. There were no more cities, no literacy and no shining gold.

After Minoan domination of the Aegean passed into Mycenaean hands, the great amount of raw gold had been attracted to their bustling island was available to the continental Greeks and the Trojans who lived on the shores of the Mediterranean. "Golden Mycenae" and "the matter of Troy" gained eternal fame in Homer's "Iliad" and "Odyssey." The Greek epic poet wrote in the 8th century B.C., five hundred years after the Trojan War and its tragic aftermath which brought the fall of Mycenae and plunged the Aegean into a chaos of cultural darkness. As Greece emerged from the barbaric centuries of privation and ruin, literacy was reborn and Homer's tales were passed on by story tellers whose listeners were eager to hear of ancient heroes. They hungered for stories of high adventure whose characters had passions and valor greater than those of mortal men. Homer and the other Greeks of the Classical rebirth crafted their tales from surviving fragments of dimly remembered events. They sung of battles and love won and betrayed, giving them vivid color with contemporary details.

The Trojan adventure recounted in the "Iliad" is the story of the incomparable Helen who was stolen from her husband Menelaus by Paris, son of the Trojan King Priam. Menelaeus' brother Agamemnon is the "king of Mycenae rich in gold" and leads the Achaeans away from their land and women for ten long years during which they lay siege to Troy. Homer detailed the golden luxury of Troy and the wealth of the royal house where palace smiths forged golden armor for the doomed Priam and the most splendid jewelry for the beautiful Helen. References to gold are everywhere in the "Iliad" and "Odyssey." The panoply of the heroic warriors gleams with gold. Their prizes are golden and when they are slain their bones are laid in golden urns. Gold is the precious stuff of solemn moments; gold tempts and betrays.

A romantic 19th-century German who had heard of Agamemnon, Achilles, Prima and Hector at his father's knee believed with all his being in the literal truth of the Homeric stories. Heinrich Schliemann ignored the scorn of scholars as he single-mindedly followed his dream, a tattered copy of the "Iliad" under his arm. Forty years of preparation culminated in a success he had never doubted and his detractors never imagined. Schliemann—merchant, war-profiteer and self-taught archaeologist—gave the world abundant, golden proof of Priam's beleaguered Troy and Agamemnon's Mycenae.

Homer's Troy we have learned in the past hundred years was one of a series of successive settlements established on a rock outcrop which commanded the narrow strait of the Dardanelles, four miles away. The first occupation was a fortified citadel which covered scarcely more than an acre of uneven ground. It was built around 3000 B.C.; the first of nine cities in which the inhabitants used some metal. A few gold trinkets and a number of tools and weapons have been found.

The second city, Schliemann's Troy, which yielded a magnificent hoard of golden treasure, was slightly larger and infinitely richer. Supplies of Trojan gold fluctuated with the swing of events in the Mediterranean. As the series of rock forts gave way to larger walled towns with palaces, Trojan goldsmiths and armourers became adept at working metal into beautiful ornaments and keen-edged, durable bronze weapons which were exported as far away as the Balkans. The Trojans had commercial contacts with the metal- and timber-rich regions around the Black Sea and with Mycenae and other Greek cities. Their cultural influence was in great demand in the war-torn centuries of the Second Millennium.

Troy dominated a fertile plain where horses were raised for export along with large numbers of goat and sheep. Thousands of spindle whorls were found by archaeologists, indicating that the Trojans also exported textiles. But the backbone of Trojan might and wealth was not agriculture but the control of the strategic narrows which divided the Mediterranean from the Sea of Marmara and the Black Sea just beyond it. The hilltop city collected tolls from ships they piloted through the straits, which the ancients called Hellespont. Troy also controlled a vital land route on the Asian seaboard and exacted fees for transporting cargoes overland to a place where they were reloaded to sail through the straits.

The second city of Troy, which Schliemann attributed to Priam, came earlier by about a thousand years, dating from the second half of the Third Millennium B.C. It was a viable city, economically prosperous with a wealthy citizenry. The personal treasures of goldwork and jewelry, which these people hid as an enemy sacked and burned the city, was what Schliemann discovered, trenching somewhat ruthlessly through the seven cities which lay above it. Without knowing it, he dug right through the actual Troy of Priam and Helen and later admitted that in his quest for the fabled Troy he "was forced to demolish many interesting ruins."

But Schliemann, small, intense and not a very sympathetic character, was not the egomaniacal treasure hunter blinded by gold lust that he has so often been made out to be. His life in which there was no time for lightheartedness or humor was a mission. He accepted Homer and other classical writers literally and consequently made some grave errors in chronology and interpretation. Nevertheless, he made signal contributions to history and archaeology. In addition to finding cities formerly considered the product of vivid imaginations, Schliemann was the first person to recognize stratigraphy, the arrangement of layers indicating successive occupations, on a site in a Near Eastern Mound. The Excavations of Troy have provided the best picture of Bronze Age development in western Asia Minor. Despite his destruction of a great deal of what he sought to preserve, the German dedicated himself to a high standard of observation, notation and publication of his findings at Troy and Mycenae, where he found gold filled royal graves brimming with treasures of a style never seen before.

Schliemann was the son of a mother who died when he was small and a minister father who failed to practice what he preached and made a meager living. Young Heinrich, born in 1822, had a rather sad and lonely childhood. His greatest pleasure came from reading the romantic tales of the Greek heroes in a book containing pictures of the sack of Troy, a book which his father had given him before he was eight. He was apprenticed to a grocer and worked from 5 a.m. until 11 at night. He met a miller who was inclined to drink and when a bit inebriated could recite long passages in Greek from the "Iliad" and "Odyssey," The grocer's apprentice didn't understand a word of the ancient language hut was fascinated by the majestic rhythms of the epic poetry. His father, his teachers, everyone had told him it was all a fancy but Schliemann was positive Troy lay waiting for him just as he had seen it in the illustration in Jerrer's "Universal History."

Unhappy at home, Schliemann ran away and found a menial job in Amsterdam. Soon after, he embarked on a ship sailing for New York, which shipwrecked in the icy North Sea. Schliemann almost died from exposure but recovered and found employment as a clerk with a firm of Amsterdam indigo merchants. He had natural business sense and did so well that the company sent him to St. Petersburg to represent them. He saw great opportunities for profit in Czarist Russia and set up for himself. He married a Russian and made close to a million dollars.

But never for a moment did he lose sight of his goal—the gleaming Homeric city. Before he was thirty the disciplined German had taught himself seven languages, four of which he became fluent in after studying each for a mere six weeks, always late at night after his business day was completed. His brother was in California in 1849 where gold had just been discovered. Schliemann made the arduous journey to seek his brother and stayed a year in San Francisco to add to his growing fortune. He began banking, buying gold from prospectors who had struck it rich in the Sierra Nevada. He departed with more than golden profits. While he was in San Francisco, California was admitted to the Union and along with everyone else in the territory Schliemann became a citizen of the United States. Even as he learned nine more languages, traveled widely and studied assiduously in preparation for the Trojan quest, Schliemann added to his enormous fortune with one clever enterprise after another. He made millions by shrewd profiteering in the Crimean War and retired from all business activities at the age of 41 to devote himself entirely to finding the sites of the Homeric cities.

One of his first steps was to take a young and beautiful Greek girl as his wife—his latter day "Helen"—who became his faithful and competent assistant. The manner in which he selected his bride was characteristic. His first marriage had been filled with friction. He was a driven man with no time for the pleasures of courting. He wrote to the Archbishop of Athens, a friend, requesting that the prelate find him a beautiful, well educated but not wealthy young woman who must, above all, be enthusiastic about Homer and the history of ancient Greece. The Archbishop found 18-year-old Sophia who proved to be a perfect match for the magnate-archaeologist.

In 1871 the two went to Hissarlik, a large mound on the Turkish shore of the Aegean where Schliemann felt sure Troy lay buried. In four campaigns of excavation in the years up to 1890 Schliemann and his "Helen" dug at Hissarlik with a large crew of laborers and uncovered the remains of not one Troy but nine. The great find which repaid his years of unshakeable faith and amazing effort came on a sweltering June morning in 1873—just one day before the season's digging was to end. Over 230,000 cubic meters of dirt, ashes, and rubble had been removed from the mound when Schliemann saw the glint of gold which led to a breathtaking treasure lying among the charred ruins of the second stratum.

The husband and wife team anticipated a routine day as they dug at some distance from the Turkish laborers. Schliemann was working at the base of some overhanging rubble which threatened to fall at any minute. His heart leaped as his eye caught the unmistakable shimmer of yellow metal. No one else noticed. He motioned to Sophia and told her to announce a holiday. She was to say it was his birthday and they could have the day off to celebrate. The crew was delighted and rushed off without questioning the boss' motive.

Schliemann and his wife dug frantically among the rocks, fearless of the overhanging stones, scooping up piece after piece of gold which they piled on Sophia's outspread red shawl and then took to their wooden cabin nearby. That night by the light of an oil lamp they inspected the dazzling hoard with dazed expressions. He never doubted that the 8,763 pieces of wrought gold jewelry and ornaments were Priam's own treasure lost during the sack of Troy by the Achaeans. He contemplated the rescued treasure, caressing the exquisite objects untouched by 4,000 years of time. He lifted an elaborate gold headdress made of 16,353 individual elements and a pair of dangling earrings and reverently adorned his Greek wife.

There were gold diadems, pendants, and brooches. There were necklaces of beads, rings, bracelets, chains, plates, vessels, earrings, buttons and even gold drawn into thread. The gold found that hot day and other hoards unearthed later showed that the Trojans, when they had gold, worked it on a magnificent scale. There is a strong stylistic and technical relationship between Troy's gold and the grave gold found in other areas of Anatolia which date from the same general period such as the Alaca Huyuk burials and far richer treasure from Dorak in Western Turkey, south of the Sea of Marmara, which mysteriously vanished not long after it was discovered. The work of the Mycenaean and Minoan goldsmiths also influenced the Trojans who translated Asiatic and Mediterranean styles into a highly personal expression of delicacy and lightness.

The gold of Troy came from several sources. Gold from concentrations worked in the Caucasus in the Third Millennium B.C. had been crafted into jewelry and vessels for individuals such as those buried at Maikop in the Kuban Valley. During the Second Millennium it is likely Troy received gold from the Caucasus and some washed from alluvial deposits on the Phasis River south of the Caucasus on the Black Sea. Some gold in the form of manufactured objects came from Crete and some raw gold must have come from the gold mines at Abydos, less than 30 miles away, which the Roman geographer Strabo visited around the time of Christ.

Schliemann's Trojan hoard made world headlines. Armchair historians who had scoffed at the daft German and his impossible quest were silenced. But poor Schliemann reaped a harvest of trouble with his treasure. He had been digging without a permit when the Turkish government heard of his stupendous find, and they demanded he turn over the treasure. His cabin was searched and not a single piece of gold was found. Sophia's family had helped him smuggle the entire hoard to Athens. Schliemann feuded with the Turkish government for years, determined that the treasure be preserved intact not for his personal gain but for history. The intellectual establishment could not deny that Schliemann had made a most remarkable find but they still regarded him with mistrust and provided little support. The Turks demanded he turn over half of what he had found. When he refused, they sued. Schliemann lost the case in 1875 and was fined 10,000 francs. He displayed the shrewdness that had made him such a giant in business when he didn't pay the fine but made the Turkish government an outright gift of 50,000 francs. He kept the treasure, worth a million francs, and the Turkish government gave him permission to continue.

The shining hoard he continued to call "Priam's Treasure" was displayed throughout Europe, and a dozen countries vied for the privilege of providing a permanent display. Germany, Schliemann's native land, won and until the Second World War the gold was exhibited at the Berlin Prehistoric Museum. During the war it was hidden in an air raid shelter at the Berlin Zoo. Priam's treasure survived the war but not its aftermath. The Russian forces seized the gold, unique in all the world, and spirited it away behind the Iron Curtain. Not a word has been heard of it since. Schliemann would be heartbroken to know that the golden diadems of ancient kings have been melted down into anonymous bars imprisoned in some subterranean vault. At Troy there is a small museum set up by the University of Cincinnati. In its display cases are bronze weapons, spindles, pots and tools but not even a single piece of Trojan gold.

The cities erected at Troy following the destruction of Schliemann's Troy II were larger and more elaborate with monumental gates and walls, but not always as wealthy. However, there were renewed stocks of gold available to Troy during the years following the sack of Crete by the mainland Greeks who took over the reins of sea trade. After centuries of scarcity the precious metals again figured prominently in Trojan culture until around 1200 B.C. when the city was sacked yet another time. Some scholars think the invaders may have been the Achaeans after they abducted Helen.

The Mycenaean civilization consumed the lion's share of gold in the few centuries of their flowering, synonymous with golden luxury. They were the middlemen between the trade of barbarian Europe and Western Asia. The scale of Mycenaean trading contacts during their heyday is documented by finds of Greek products in distant lands. Lovely Mycenaean ceramic ware, metal objects and other artifacts have turned up occasionally in Europe and with amazing frequency in the zone stretching from Italy to the Levant and south into Egypt. They dominated the Cycladic Islands, Rhodes, Crete and colonized Cyprus. They had overcome the innate unease a farmer feels for the sea to become highly proficient mariners.

Mycenaean civilization developed around 2000 B.C. as settlements of early Helladic, or Greek Bronze Age, people were leavened by immigrants from Asia Minor. These invaders attacked communities then often remained. They were skilled horsemen who brought with them a potent weapon, the horse drawn chariot, which they introduced to European warfare. The mainland communities prospered. By 1700 B.C. they were already shipping some agricultural products to Egypt; perhaps aboard Minoan carriers. In return for their excellent olive oil, wine, hides, timber and dye they received gold, linen, papyrus and rope. Contact with the Minoans fostered Mycenaean cultural development. The exposure to the Cretans was visible in almost every craft—goldwork, fresco painting, pottery and carved seals were all influenced by the superior Minoan culture.

Mainland centers of wealth and power developed gradually. Around 1600 B.C. the first to emerge, and the greatest of all, was

Mycenae in the eastern Peleponese. The accomplished and enterprising people of eastern and southern Greece who shared a language and culture referred to themselves as Achaeans. This is what the Classical Greeks called them in the legends which immortalized their achievements, along with their gold and prowess at war. The inhabitants of the mighty city of Mycenae have given their name to the Late Bronze Age culture which laid the foundations for European civilization.

The Mycenaeans established thriving metal industries and the bulk of their exports was no longer agricultural. Trading bronze weapons—the magical double-axes they had incorporated into their own culture—glass and some gold, the Mycenaean ships called at ports where they could find gold, essential metals and the exotic luxuries that flourishing cultures crave. The increased Mediterranean demand for gold, copper, tin and Baltic amber quickened the previously sluggish progress of development in much of Europe. The Mycenaeans opened up the western end of the Mediterranean as they sought metals, particularly the tin needed for their bronze manufacturing. Cypriote copper ingots which were exchanged as barter currency have turned up in Sardinia, Malta, Romania and France.

Aegean mariners sailed to southeast Spain and settled traders in Sicily and perhaps southern Italy as well. Some enterprising Mycenaean traders settled in ports of Syria, Lebanon and Israel where they exchanged Greek goods for ivory, spices and purple dye. Their presence was felt all over Europe and the Near East and had an important influence on Late Bronze Age technology and economics. Artifacts of gold, bronze and ceramic of alleged Mycenaean origin have been unearthed in Great Britain. It is possible they traveled that far pursuing the promise of Irish gold, tin and amber traded on those shores by the Minoans before them. It also seems possible that the Mycenaean goods traded so far north were carried by Phoenician merchant ships from Byblos.

The bulk of Mycenaean gold came from Egypt. The gold of southeastern Europe probably reached Mycenae through the same

Adriatic ports the Minoans had frequented. Some came from Britain, a bit was gained in trade with the Byblos merchants. The most interesting source of Mycenaean gold was the Caucasus. Recently archaeologists working west of Tiflis in the former Soviet Union where so much gold of the nomadic Scythians was found came across gold and bronze objects of apparent Mycenaean origin which date back to the 15th century B.C. The grain of truth around which the classical legend of Jason's quest for the Golden Fleece originated may very well have been a Mycenaean voyage to trade for Caucasus gold on the Black Sea.

The greatest of the myths of the Mycenaeans is the tragedy of Agamemnon's murder and Orestes' revenge. In recent years, archaeologists working among the ruins of Agamemnon's great Bronze Age palace fortress have made discoveries indicating once again that legend can offer the answer to questions plaguing historians. It has long been known that the ca. 1200 B.C. rich city was destroyed by fire. The cause of the conflagration was popularly attributed to either a natural disaster of some kind or the ravages of invading Dorian tribesmen from the North. New evidence points to yet another cause—civil war which broke out at the end of the Trojan War just as the legend states.

The legend, which has inspired poets and dramatists from Aeschylus to Eugene O'Neil, tells of Agamemnon returning home after ten years as commander in chief of the Aechean forces against Troy. He and his retinue were treacherously murdered by his wife Clytemnestra and her lover, Aegisthus. Eight years later Agamemnon's son Orestes avenged him by killing his adulterous mother and Aegisthus.

Excavations carried out by the Greek Archaeological Society from 1958 to 1972 proved that Agamemnon's city was economically prosperous and politically strong up to the end of 1200 B.C. Then it appears Mycenae was destroyed in the intericene war that followed the royal murders. Sporadic raids by the Dorians followed and prevented Mycenae from resuming its supremacy. The decline and chaos culminated in the capitulation of the ru-

(Above, left) Golden mask recovered at Troy by Heinrich Schliemann. *(Above, right)* Ram in the thicket gold statuette from Ur. Ca. 2600 B.C.

(Above) Mycenaean Gold goblet 1600-1550 B.C.

(Above) Rillaton gold cup resembles a late Neolithic ceramic beaker with corded decoration but dates to a much later period of ca. 1650-1400 B.C. The cup is of note due to its Aegean style metalwork of the period and resembles similar finds from the Greek site of Mycenae, suggesting cultural and trading links with the Eastern Mediterranean.

Chinese gold bracelet 18th Cent. A.D.

(Above, left) Silver gift cup Ludwig Krug 1490-1532. *(Above, right)* Ram in the thicket gold statuette from Ur, ca. 2600 B.C. *(Lower left)* Inca Portrait Beaker 1000–1600 A.D.

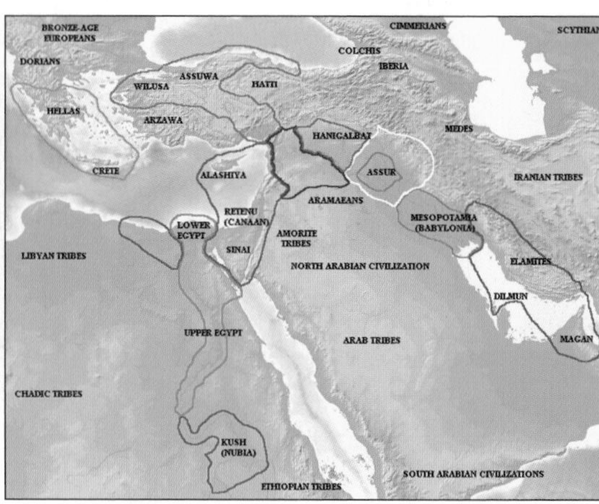

(Right) Map showing cultural distribution, ca. 500 B.C.

Chapter 6

Biblical and Phoenician

The surprising Aegean discoveries were not the last to open new and important chapters in ancient history and cast light on the role of gold in antiquity. Archeologists have continued to inch their way through and sift the dusty rubble at countless sites. From time to time their patience pays off in exciting discoveries that push back the hazy frontiers of prehistory.

In the 1970s a young Italian archaeologist working in northern Syria made a really sensational find. Paolo Matthiae's revelations of the existence in the Third Millennium B.C. of the kingdom of Ebla, which rivaled those of Egypt and Mesopotamia in sophistication and complexity, is being hailed as the greatest archaeological breakthrough of the generation, possibly of the century. Among other things it gives a new dimension to the movement and use of gold in the 150 years between 2400 and 2250 B.C. when Ebla flourished as an important cultural and commercial center. This new fund of knowledge is a reminder that the future can be expected to divulge even more of the past—the final line has not yet been written on any aspects of antiquity.

Matthiae and his team from the University of Rome exhibited a dedication Schliemann would have admired. They toiled a dozen years in the barren wilds of Syria excavating a vast mound of dirt about thirty miles south of modern Aleppo. The high mound covers more than 140 acres in a zone historically dismissed as a desolate wasteland without culture or commerce and only fitfully inhabited in ancient times by migratory herdsmen. Matthiae dug on despite the experts' skepticism. In 1975 his persistence was richly rewarded with the discovery of 15,000 clay tablets. They are inscribed with cuneiform script chronicling the history of the previously unknown civilization that held sway over a large area more than 4,300 years ago.

The tablets were stored on wooden shelves of the archives of what turned out to he the royal palace of the lost kingdom of Ebla, a powerful gold-trading state whose sphere of economic and political influence stretched from the Sinai on the south through modern Israel, Lebanon and Syria to Cyprus on the west and the highlands of Mesopotamia on the east. The capital city alone had 260,000 inhabitants.

Ebla was a trading nation involved in a complex commerce in processed metals and textiles. Imported gold, silver, copper and tin were turned by Eblaite smiths into manufactured articles in precious metals, bronze and copper for export to neighboring Mid-Eastern states. They also engaged in a constant struggle with the Akkadians of Mesopotamia whose king, Naram-Sin, eventually destroyed the royal palace where the tablets were unearthed. However, there were periods when the Eblaites dominated their arch rivals and were powerful enough to conquer cities in the Euphrates Valley, stripping them of their gold and silver treasures. One tablet details a successful campaign against the kingdom of Mari. The conquered king sent tribute to Ebla of 880 pounds of gold and 11,000 pounds of silver.

The records are written in the most ancient Western Semitic language known, which is similar to the biblical Hebrew spoken centuries later. The Eblaites borrowed the Sumerian's cuneiform characters to record their language and a schoolboy had to learn both languages. The discovery of the earliest known bilingual vocabularies in the archives was the key to deciphering the tablets. The most exciting facet of the find is the evidence suggesting the existence of the places and people mentioned in the Old Testament. Less than ten percent of the tablets thus far have been translated and they contain many intriguing parallels between Eblaite culture and that of the Hebrews. Just as Schliemann's finds vindicated his ancient sources, this new find is divulging the substance of biblical legend. The remaining tablets, which will take years to puzzle out, promise much more information to substantiate Old Testament references and fill in gaps in the knowledge of Bronze Age gold.

The dynastic kings Of Ebla were anointed with oil at their coronations as were David and Solomon. Moreover the tablets list Eblaites whose names translate into David, Israel, Saul, Abraham and Esau and refer to a place called "Urusalima" which could well he the earliest known reference to Jerusalem.

Most intriguing of all are the frequent references to the illustrious Eblaite monarch Ebrum. His name bears a startling resemblance to the "Eber" listed in the Genesis genealogies as the distant ancestor of the Hebrew patriarch Abraham who led his tribe out of Ur to settle in Canaan some three centuries after the ceding of the Eblaite Empire.

The Hebrews and the Phoenicians, their Semitic neighbors on the eastern Mediterranean seaboard, were the world's principal gold dealers in the millennium before Christ. There are over 420 references to gold in the Old Testament which indicate the metal was well known to the Hebrews and highly prized. Hebrew has several words for gold. The most common is zahav which occurs 385 times. A poetic term for gold was *ketem*, which was probably borrowed from the Egyptian language. There were words to indicate gold's color and purity, dust, ingots, leaf, plate and more.

To the Hebrews gold stood for the highest good and perfection of spirit as in the glorious city of Pure Gold, the Holy City of Jerusalem described in the New Testament. It represented power and majesty, but more often gold is the symbol of decadence, evil and a falling away from the good. As when the Israelites, impatient for the return of Moses from the mountain, demanded Aaron make them a golden god they might worship and they turned from Yahweh to the golden calf idol. When the patriarch returned and saw what had happened, he ground the calf to powder. He then mixed the powdered gold with water and made the Israelites drink it—the aurum potabile so beloved of medicine men through the ages.

Frequent exhortations in the Old Testament not to trust or worship gold, still linked in men's minds with the sun. Yahweh ordered his people to make no idols of gold nor worship any precious metals. Job recalled the antecedent sun worship and the gold-sun association when he declared, "If I have out my faith in gold and my trust in the gold of Nubia... if I ever looked on the sun in splendor or the moon moving in her glory, and raised my hand in homage, this would have been an offense before the law, so I should have been unfaithful to God on high."

Biblical references mirror the ambivalent human responses to gold that are as old as man's knowledge of the precious substance. There are admonitions against coveting gold, prohibitions against using it on the one hand but on the other gold is deemed most excellent for the embellishment of the Temple and all manner of ceremonial objects.

The Old Testament is an excellent and reliable source of information about Hebrew activities in prospecting, mining, trading for gold, tribute-collecting and acquiring gold in war. It also details the lavish use of gold in Solomon's reign when the Temple and the great royal place were constructed with Phoenician help and filled with gold from distant lands. The work of goldsmiths is described from refining and assaying to finishing various types of gold objects. Several goldsmiths are named in the Old Testament and the goldsmiths of Jerusalem all belonged to a guild.

In the centuries following the collapse of the Aegean civilization the golden spotlight shone again on the Near East, but not this time on the Egyptians whose empire was in decline. The depredations of the Sea Peoples had further weakened the might of the Nile kingdom, which had dangerously overextended its imperial reach and was unable to maintain far-flung frontiers. The Semites with a long tradition as traders took advantage of their favored location to become the focus of the commerce and the use of gold before the Classical Greek period.

The abundant gold of the Bible is listed as coming from the magic golden lands of Nubia, Havilah, Sheba, Cush, Tarshish, Reemah and especially Ophir. These semi-legendary lands have shimmered mirage-like before the dreamy eyes of gold-seekers for thousands of years. They are mentioned in the tenth chapter of Genesis, which also lists a king named Eber, perhaps the Ebrum of Ebla, as the great-great-grandson of Noah. The chapter is a geographical and genealogical table which lists Noah's descendants who founded the separate nations of the earth following the flood.

According to the genealogy Eber's son Joktan was the father of the gold countries of Sheba, Ophir and Havilah. Gold is the first element and the first metal mentioned in the Bible. In the second chapter of Genesis a river flowed out of Eden to water the garden and there it divided into four streams. The name of the first is Pishon; that is "the river that encircles all the land of Havilah where there is gold and the gold of that land is good."

Very soon after Adam and Eve were driven out of the earthly paradise men began to "gather gold, silver, gems and pearls from all parts of the earth." The precious metal for the Golden Calf, for the heavy bracelets which adorned Rebekah and for the 80 pieces of gold traders paid for Joseph came from war booty and trade, the gold for his crown and the lavish appointment of Solomon's Temple and palace. Where did the gold come from? How did the Hebrews rise from rough tribesmen, wandering in the desert with their flocks and herds, to full-fledged nationhood crowned with the golden trappings of wealth and power? Tracing the rise of the Hebrews and their cousins, the seafaring Phoenicians, is a fascinating exercise that illustrates the story of gold in the First Millennium.

It is the story of trade and the greatest mercantile people of antiquity; the Phoenicians and their descendants the Carthaginians whose prowess as shipbuilders, sailors, navigators, commercial middlemen and manufacturers was unexcelled. The Phoenicians were also miners and metalworkers who sailed farther into the unknown than any of their predecessors to exploit new sources of gold that the world craved. Their closely guarded secrets unfortunately died with them and neither the Greeks nor the Romans ventured far on the Terrible Ocean or the Green Sea of Darkness, as the Atlantic was known.

Searching for gold, they traded with their enemies the Greeks, with Africa, Iberia, Britain and overland with Transylvania and Bohemia where they bartered the gold and silver from one place for the utilitarian metals from another. Supplies of Central European gold, which were no longer tapped by the Mycenaeans, added to the world's growing stock. More and more people who were neither priests nor kings aspired to have a gold ornament. Gold rings which had been used in early Egypt, in the early Irish gold culture and to some extent in Troy and Mycenae, begin to figure in Levantine life. "Thy cheeks are comely with rows of jewels, thy neck with chains of gold." The Song of Solomon mentions gold chains which, like rings, served as adornments and a particularly attractive form of portable currency. Rebeccah at the well was presented by Abraham's servant with a gold nose ring of half a shekel's weight and bracelets weighing ten shekels.

The Phoenicians origins are less clear than those of the Hebrews but both evolved from wandering desert tribes that filtered to the hospitable fringes of the Fertile Crescent, the great semi-circle of arable land which lies like a horseshoe open to the Arabian Desert with mountains bordering its outer rim. The area of modern Israel, Lebanon and Syria was under almost continual migration and invasions in the Bronze Age. As early as 3000 B.C. people who called themselves "Canaanites" or "merchants," but whom others called Phoenicians, had settled on a cliff overlooking the Mediterranean and founded a bustling port and trading center known as Cubla to the Assyrians and Babylonians, and Byblos in Greco-Roman times. The merchants of Byblos represented a mixture of the original coastal inhabitants and the Semitic-speaking immigrants. The mountain slopes in back of Byblos provided the cedars of Lebanon and pine, which was their first trade item. Egypt was receiving wood and stone for buildings and cedar resin for incense from Byblos at the beginning of the Third Millennium. Egyptian tomb reliefs ca. 2500 B.C. show stubby merchant ships from Byblos unloading on the Nile. Profits furnished the tombs of Byblos' princes with magnificent funerary objects of gold, silver, ivory, semi-precious stones and rare woods.

The shipbuilding merchants of Byblos also dealt in the precious perfumes, oils and incenses of Arabia which were consumed in great quantities. The demand for fragrant frankincense and aromatic myrrh always exceeded demand and a great profit was made by those intrepid men who led long caravans of donkeys and later camels from the Hadramat wilderness of southern Arabia to the coast.

Byblos, lying between Egypt and the Euphrates, was a also a vital link for the Mesopotamians. As the great empires grew, they required more metals and Byblos and the other Canaanite cities figured prominently in the metal trade. By the second millennium, a number of city-states were operating in the sea trade. More people had wandered in from the interior and mingled with the native population. Later, Aegeans fleeing northern invasions, came with their sea lore. Gradually these mixed peoples became the bold shrewd Phoenicians called by Homer "a race of mariners."

With the fall of the Egyptian Empire after 1200 B.C., the ships of Egypt in the Eastern Mediterranean had disappeared. The same fate overtook the fleets of the Aegeans at about the same time. Thus the Eastern Mediterranean was left unoccupied by merchant fleets and by 1000 B.C. the Phoenician cities were taking advantage of this opportunity

It was then that Tyre surpassed Byblos as the major seaport and retained this position until being destroyed by Alexander the Great in 332 B.C. Ezekiel said about Tyre, "Clever and shrewd as you are, you have amassed wealth for yourself, you have accumulated gold and silver in your treasuries, by great cleverness in your trading you have heaped up riches and with your riches your arrogance has grown." With Tyre and Byblos the other principal city-kingdoms of Sidon, Arad and Berytos (modern Beirut) shared a common culture, religion and language but competed with one another in trade. Profit rather than conquest was the goal of these exceptionally adventurous people who developed the alphabet and the abacus to facilitate their commerce. The Phoenicians learned astronomy from the Babylonians and used their knowledge of the stars to guide them on the trackless seas. Phoenician enclaves grew up at many of the Mediterranean's natural, easily defensible harbors. They settled at Carthage in the 9th century B.C. and in the next century at Sicily, Tunisia, Sardinia, east Cilicia and even the Taurus Mountains. Soon after, they also dominated Cyprus and Malta.

For centuries the Phoenicians and the Carthaginians controlled ancient commerce. The trade north and south, east and west, all passed through their clever hands. The canny sea kings maneuvered to survive Egyptian, Assyrian, Babylonian and Persian assaults which greatly reduced their economic power. Finally, Alexander the Great laid siege to Tyre in 332 B.C., still the principal Phoenician stronghold. After holding out for nine months, the island city which had withstood a 13-year siege by Nebuchadnezzar the Second in the 6th century B.C., fell to the world conqueror, and the Phoenician mariners were eclipsed by the Carthaginians whose control of the Western Mediterranean had already siphoned off much of their trade. Carthage was Rome's most dangerous rival. Carthaginians, during the 7th to 5th centuries had established colonies in Algeria, the Balearic Islands, southern Spain and along the Moroccan coast. Sailing from these ports, they traded for gold on the west coast of Africa and tin in Cornwall. The search for raw materials led them to the Azores and very possibly North and South America where bits of archaeological evidence to confirm this possibility are being pieced together.

Everywhere they went the black-bearded, keen-eyed merchants spread products, inventions, customs and population. The Phoenicians were so tight-lipped about their monopolistic control of raw materials that they made no charts and never committed any of their vast lore of geography and navigation to writing. What they offered were spine-chilling tales of ghastly demons, treacherous tides and entangling seaweeds that lay in wait for any who ventured out into the *Mare Altum*. More than 20 centuries later as the Portuguese gingerly felt their way down the west coast of Africa, belief in the Phoenician chimeras was still current. More than once a Greek or Roman ship shadowing a Tyrian or Sidonian galley was lured onto rocks or led on a wild goose chase.

From initially acting as middlemen the sea-traders became skilled craftsmen and metallurgists. By selling their own wares, they greatly increased their profits. They were artistic magpies incorporating the technique and motifs of various peoples in their work. Trading cheap, mass-produced trinkets for far more valuable gold and raw materials was a Phoenician specialty. The Punic traders originated wholesale trade in cheap goods. Costume jewelry, assembly line votive figurines, inferior perfumes, combs, crude glass ornaments and painted ostrich eggs were their staples But they also built their trade on woods, the famed purple dye called "Tyrian purple," fine dyed and embroidered textiles, jewelry, semi-precious stones and carved furniture—veneered with rare woods and embellished with inlays of gold, ivory and gem stones.

The Phoenician goldsmiths fused divergent elements into objects crafted to satisfy current market demands. They evidently felt little need to express a distinctive individual style. The Phoenicians picked up elements of Mycenaean and Minoan goldwork. From the Egyptians they learned enameling, granulation and repoussé. With equal ease the goldsmiths of Tyre and Sidon borrowed Assyrian, Cypriote and Mesopotamian motifs blending them in a mongrel style. A Phoenician necklace or golden cup reflects the many currents running through the Levant. In time, the oriental emphasis was moderated by exposure to the Hellenistic world.

Master goldsmiths from the Phoenician kingdoms set up shop in Greek cities starting in the 8th Century B.C. They trained apprentices and greatly influenced Greek jewelry of the so-called "orientalizing period." Homer describes Phoenician goldsmithing in a number of passages, praising the magnificently fashioned Sidonian bowls which had scenes worked in relief. The Greeks, who otherwise loathed the Phoenicians, reluctantly granted them first place among goldsmiths. Recent excavations of rich burial sites in Lebanon have supplied a wealth of exquisite gold artifacts that, in spite of a certain promiscuity of style, the Phoenician goldsmiths were superbly skilled.

To the south of Tyre in the area historically known as Palestine, another group of Semitic traders developed over the centuries. They were the Hebrews and related tribes whose forefathers had roamed the sandy wastes from oasis to oasis since earliest times. They followed their flocks and herds to desert fringes that turned briefly green with tender grasses following scant winter rains. From time to time these people gained a foothold in areas previously settled by the Canaanites who built stoutly walled towns at the western end of the Fertile Crescent.

Palestine was a narrow strip about 150 miles long between the mountains and the sea which had been inhabited for some 200,000 years. During the Stone Age the Jordan Valley, the world's lowest area at 1,300 feet below sea level, was the site of the most advanced culture in the world judging from the excavations at Jericho. But the high level of development was not sustained in subsequent ages, for much of Palestine was unproductive. In the southern end the Sinai desert intrudes on land making it too arid for cultivation on an intensive scale. The central zone was composed of rugged limestone hills. Only the northern valleys offered prime arable land. There were few natural resources save salt and the copper of the Negev and the Sinai where ruined workings of the world's oldest sophisticated underground mines have been found in the hills near the port of Eilat. Archaeologists aided by mining engineers have explored a network of 200 vertical shafts and galleries in the Timna Valley which Egyptian miners dug for copper ore with stone hammers and bronze chisels beginning in the 15th century B.C., and seem to have accidentally discovered the secret of smelting iron in the process of refining copper ore.

Palestine had harbors only at its northern end, but these in the firm grip of the Phoenicians left the Hebrews no access to the Mediterranean. Moreover, tiny Palestine was terribly vulnerable. Open to the ravages of invaders arriving by sea, Palestine was swept by the Asiatic Hyksos who proceeded on to conquer northern Egypt. A couple of centuries later these barbarians were back, this time fleeing before the Egyptians who managed to absorb Palestine into their empire. Lying between the empires of Egypt and Mesopotamia, Palestine like the Phoenician states was frequently subject to foreign domination, caught in the pincers of the giant powers.

Abraham arrived before the Hyksos invasion, and Moses led his people out of bondage in Egypt to settle in Palestine about the 14th century B.C. These Semitic nomads echoed much of the culture of the settled peoples whose towns they captured. The first king of the federated Hebrew tribes was Saul. One of his warriors named David, a man of the tent dwellers of southern Palestine whose people clung to their ancient ways, became king of the south. He was able to unite his kingdom with the more urbanized and prosperous north. Because powerful neighboring states were momentarily distracted, the wise king succeeded in greatly extending his dominions. From his capital at Jerusalem the poet warrior-king ruled wisely and well, laying the foundations for his son Solomon's grandiose gold-seeking expeditions and construction projects, which were undertaken with the aid of David's friend, King Hiram of Phoenician Tyre. David's conquests gave Israel temporary control of important trade routes. A great part of his wealth came from a transportation tax and another upon "the traffic of the spice merchants and all the kings of Arabia."

Unfortunately Solomon's opulent indulgences and ambitious projects overtaxed his people's resources. At his death the united kingdom his father had welded ruptured as the ten northern tribes seceded to form Israel, which was conquered by the Assyrians who took Samaria, the capital, in 722 B.C. The southern kingdom of Judah, from which the word Jew may derive, lasted until Nebuchadnezzar destroyed Jerusalem in 587 B.C. and took all of the gold treasures of the temple and palaces and the elite into exile in Babylon.

Palestine enjoyed a golden age under Solomon. His reign was the culmination of Israelite political history. When he came to the throne, ancient Palestine had been transformed from a completely agrarian country into an economically viable nation involved in international trade. Without access to the sea, great rivers or bountiful natural resources, Palestine's saving grace was its strategic position. The chief caravan routes from Egypt to Syria and from the Mediterranean to the hills beyond the Jordan River passed through the country. Camel transport was introduced about 1000 B.C., replacing the previously slow donkey caravans that had been limited by requirements for frequent water. The treasures of Arabia Felix—perfume, incenses and gold were no longer dependent on the 1,200-mile-long incense route which wound through the sandy wastes from oasis to oasis. Some of the camel trails had their terminus in Israel. Solomon's government collected tolls and levied customs on all traffic. King Solomon also taxed his own people oppressively and received golden tribute from the Kings of Arabia and regional governors.

All this brought him gold…a lot of it. But it was not enough for the king who wished to replace the portable tent which the Hebrews had always used as a temple with the most magnificent house of worship the world had ever seen. To the north his friend, King Hiram of Tyre, also craved more gold even though his island trading kingdom was so rich the Bible described it as a place where silver was heaped up as dust and fine gold as mire of the streets.

The two monarchs formed a joint trading expedition to bring back the fabled gold of Ophir. In one of history's most interesting partnerships the two Semitic kings built a fleet which sailed periodically from the modern port of Aqaba and returned with about 34 tons of gold, worth more than 600 million dollars today. Hiram had aided Solomon's father. Indeed, the Phoenicians were accustomed to building ships, temples and even cities for other countries. They courted foreign rulers with rich gifts—including gold—to reduce the possibility of invasion and to entice potential customers. Assyria's king, who had received such tribute from the kings of eight of the sea cities, was so pleased with the quality of the gifts that he ordered Phoenician materials and workmen to make furnishings and metalwork for his vast new royal palace.

David had accepted gifts from Tyre and planned to use Phoenician workmen to build a temple to the one god. He had pledged his personal fortune of some 3,000 talents of gold to finance the project which he had long dreamed of. In addition he taxed his people an additional 5,000 talents. He died before the project got underway but left detailed plans to his son. The pages of the Old Testament record every detail of the planning, construction and furnishings of the sacred building around which the religious and political life of the Jews revolved for so many generations.

Solomon approached Hiram because the Hebrews needed timber, dressed stone, rare woods, sumptuous cloths and lots of gold, silver and bronze. They also needed the skilled artisans to execute such a grandiose scheme. The merchant king arranged to send Solomon a team of architects, carpenters, masons and gifted smiths in gold and bronze as well as the necessary raw materials. In return, during the seven years the Temple was under construction, Hiram received a huge annual payment in wheat and olive oil to sustain the growing population of Tyre. Not since the Pyramids had there been construction on such a scale. Solomon conscripted labor armies. Some men worked at home, others went to Lebanon. Thirty thousand woodcutters were sent in relay teams of 10,000 a month to fell the towering firs and cedars of Lebanon. The timber was brought to Jerusalem by 70,000 porters who also transported the stone quarried and dressed by 80,000 masons.

The monument to the Hebrew god was built like a Phoenician temple, a narrow building approximately 135 feet long by 35 feet wide and 50 feet from floor to ceiling with walls ten feet thick and entered by a flight of steps. The three interior rooms were ablaze with gold. "He covered the whole house with gold, its rafters and frames, its walls and doors, and the floor too was overlaid with gold." The walls were hung with sheet gold embossed with patterns of palm trees, lotus blossoms and chains. Gold was used everywhere. The inner sanctuary was sheathed in 600 talents of fine gold and even the nails were golden. Altars, tables, enormous stands for oil lamps were crafted by goldsmiths who made the golden ark which held the stone tablets of the commandments, and was guarded by two great gilded sphinxes with outstretched wings, called cherubim in the Old Testament. Solomon ordered all manner of gold liturgical objects including bowls, cups, saucers, snuffers and firepans. Solomon also had 200 shields made of beaten gold worth 600 shekels apiece and 300 bucklers of beaten gold worth 300 shekels apiece. Solomon finally brought the gold and silver treasures of his father and added them to the inventory of the Temple treasury.

The gold from Ophir was also used for the sumptuous palace Solomon, a "great lover of women," built for his family, which included a reported 700 wives and 300 concubines. Each of the wings was four times the size of the temple. The growth of luxury during his time as described in the scriptures indicates there was no acute shortage of gold among the princes of the people whose forefathers had roamed the desert. The palace took 13 years to build. From the throne to the tableware only gold was used for "silver was reckoned of no value in the days of Solomon" whose oriental love of splendor seemed boundless.

How did Solomon get the gold of Ophir? His part of the bargain with Hiram was to furnish the Red Sea port of Ezion-Geber on the Gulf of Aqaba near modern Eilat. But where was Ophir? Some scholars have equated Ophir with the Egyptian Punt, where Ramses III mined for gold a few centuries previously. It is also thought that because Solomon was married to the Pharaoh's daughter, he may have had inside knowledge of Ophir's location.

Not long after the days of Solomon, the location of Ophir became lost and men searched for it in many areas of Africa. Some contended it lay at the mouth of the Indus in Hindustan. Josephus, the ancient Jewish historian, and Jerome, the 4th century A.D. Christian scholar, both thought the gold of Ophir came from the port of Supara north of Bombay. Arabian sailors had probably learned to take advantage of the monsoon to sail southeast in summer across the Indian Ocean, returning in winter by the northwest monsoon long before the 1st century B.C. when the Greek pilot Hippalus is credited with this discovery.

In 1505 a Portuguese squadron sailing for the fabled Indies dropped anchor at a small African port named Sofala on the Mozambique Channel. They saw that Arabs, or Moors, as they called them, brought gold to the coast from the interior and thought they had found Ophir. For a time the Portuguese also sought Ophir in the Golden Chersonese of the Malayan Peninsula. The Mashona and Matabele region between the lower Zambezi and the Limpopo rivers of Rhodesia remained a favorite site. The mysterious monumental stone ruins of Zimbabwe were associated with the Golden Empire of Monomotapa, the object of many gold-seeking expeditions in the Age of Discovery. The ruins lie among gold diggings that geologists figure produced as much as 25 million fine ounces before the late 19th century. However, the citadel and cyclopean walls turn out to have been built centuries after Solomon.

The 16th century was marked by the dream quest of the Spanish in the Western Hemisphere. They hunted the golden chimeras such as El Dorado, the Golden Cradle, buried cities and phantom lakes brimming with gold. The conquistador was not after gold underground as much as the accumulated treasures of generations of Indian civilizations who had buried it with their dead or stored it in their temples. When they failed to find the golden cities in the wilderness, they sought them by marching westward to the sea. When no gold sands spangled the shores, they looked to the offshore islands which Indian legends had heaped up with treasures. Looking for the Isles of Solomon, the purported site of Ophir, their frail vessels reached the Galapagos and other Pacific islands, and incredibly arrived in the East Indies in 1567, having sailed westward for 7,000 miles until they sighted land after two months.

According to the Portuguese Lopez Vaz, the archipelago which bears Solomon's name on today's maps was named because "to the end that the Spaniards, supposing them to he these Isles from whence Solomon fetched gold to adorn the Temple at Jerusalem, might he the more desirous to go and inhabit the same." However, when the Spanish Crown sent a large contingent of settlers to colonize the Solomon's in 1595, their ships were unable to find the elusive islands which had been found and described 30 years before. Dissolved into romantic legend they remained lost despite their size until Carteret and Bougainville entered the South Seas two centuries later and rediscovered them where, after all, there was no gold.

When the two Semitic monarchs became joint partners, Israel had neither ship-building materials, shipwrights, seamen nor navigators. Tyre supplied all these. In addition to the Biblical accounts of the venture there is a passage from a Phoenician source supporting the Old Testament by saying that "there were great palm forests in the neighborhood of this place but there was no timber suitable for building purposes so Joram (Hiram) had to transport the timber there on 8,000 camels. A fleet of ten ships was to be built."

The fleet set sail, probably laden with refined copper from Timna mines, with the experienced Phoenician officers and seamen. "And these in company with Solomon's servants, went to Ophir and brought back 420 talents of gold to Solomon," of which each king got a share of the first 13 tons. On another joint voyage the ships returned with 666 talents or about 21 tons. The voyage was said to take three years, one to sail each way with a layover in between. The "fleet of merchantmen came home bringing gold and silver, ivory, apes and peacocks." Some renderings insist that the cargo included slaves, which is not surprising since the Phoenicians were renowned slavers.

They were capable farmers and it was their practice on long voyages to go ashore when fall came, plant some corn and wait to harvest it before putting to sea again, a fact that may account for the long duration of their voyages to Ophir, rather than the actual sea-going voyage. Herodotus describes a commission from the Egyptian Pharaoh Necho, ca. 600 B.C., when Phoenician explorers circumnavigated Africa on a three-year voyage. They were the first men to do so and the last for another two millennia. It is perhaps the Biblical voyage of three years duration that coincided with the sail around Africa, starting from the Red Sea and returning home through the Pillars of Hercules.

The geographical location of Ophir has been a matter of intense speculation since Solomon's time. Ophir was a magic word—the richest of the golden lands—which intoxicated the ancient Jewish authors. Isaiah speaks of the "golden wedge of Ophir." The book of Job warns that wisdom cannot be valued with the gold of Ophir, and another evocative passage says, "Then shall thou lay up gold as dust, and the gold of Ophir as the stones of the brooks." The spell of Ophir was greater even than that of Havilah, Parvaim, Tarshish or the other golden lands. It beckoned gold seekers from all over the globe.

One of the most exotic episodes in the Old Testament was the state visit the queen of Sheba paid Solomon. She, like other rulers, courted him. Word of his fame as the king who outdid all on earth in wealth and wisdom reached the capital of the southwest Arabian kingdom of Saba, a dream city of gold and subtle fragrance grown incredibly rich over countless centuries in the incense and gold trade. The Sabaean merchant princes lived in houses resplendent with gold and gems steeped in oriental luxury, stories of which were told as far away as the Mediterranean. She had set out across the desert from the spice metropolis in what the Romans called *Arabia Felix*, Happy Arabia, with, a sizeable retinue and a camel caravan bearing magnificent gifts to impress Solomon—gold, precious stones and the perfumed resins, oils and incenses of the enchanted land.

Arabia Felix remained a forbidding but enchanting land long after Sheba. Greek and Roman authors wrote that gold in Arabia was not mined by men but by industrious ants. According to Herodotus they were allegedly larger than foxes but slightly smaller than the gods and more fierce than the most savage lion. They dug gold from the sand when it was cool and rested in the heat of the day. While they slept, men stole their gold. Those who didn't escape were chased by the giant ants and perished. It is true that burrowing ants and other insects and animals have acted as prospectors, digging up earth which contains gold. Miners sometimes discovered rich strikes this way. It was popularly believed that in the remote lands famous for their gold deposits that it was mined by animals, rather than man. Other giant ants were said to mine gold on the Red Sea islands. Ethiopians from one shore and Arabians from the other vied to steal this eastern gold. Africans were credited by ancient sources with a clever scheme. Each morning they rowed to the islands with brood mares fitted with a great open saddlebag. The horses' foals were left on the Ethiopian shore. While the mares grazed, the laboring ants filled the bags with gold since they seemed the perfect storage place. After the bags were filled with gleaming dust, the Ethiopians brought the hungry foals to the shore where their cries reached the mothers who immediately headed for their offspring, swimming away from the furious ants who are not recorded as ever having learned from one day to the next to repeat their folly. Camels, not mares, carried off gold dug by ants in India according to a tradition Pliny the Elder preserved in his writings. Islands of gold and silver were throught to lie at the mouth of the Indus River. The precious metals were allegedly mined there by creatures the size of wolves and with fur like cats. "Such is the speed and savagery the love of gold awakens in them" that the camels and riders are torn from limb to limb.

Gold in Arabia, wrote Agatharchides of Cnidus, who was tutor to the young Egyptian king, Ptoley Soter II at Alexandria in the late 2nd century B.C. and a voluminous author, came in solid pieces, the smallest "not less than an olive stone and the biggest a walnut." These nuggets were worn on strings as necklaces and bracelets and chunks were removed when necessary to exchange them for goods or services.

During the Emperor Tiberius' reign the Roman Strabo wrote about the camel herding Arabian nomads in an area where "there is a river which carries down gold dust, though they have not the skill to work it...gold is dug from their land, not as dust but in nuggets which need little refining...they pierce these and string them on flax, with transparent stones alternating and make chains and put them on their necks and wrists...and they market their gold cheaply to their neighbors...for they lack the skill to work it, and the metals they get in exchange, silver and copper, are rare in their country." Diodorus, writing even earlier, marveled that the river "brings down so much shining gold dust that the mud at its mouth positively glitters."

Southwest Arabia, remote home of the Sabaeans, the Hadramatites, the Qatabians and other wealthy folk was long considered the most likely home of Ophir. Alternative suggestions listed various other coastal areas and even the gold-bearing islands of the Red Sea between Egypt and Arabia. At least one author contended that the gold of Ophir was actually Hottentot gold from Rhodesia which he identified as the biblical golden Havilah country brought by Arabian trading vessels to a port midway along the southern coast of Arabia. According to this theory, the port was Ophir, perhaps the greatest of the ancient world, opposite the island of Socotra, where the merchandise of Africa, India, Arabia and the Eastern Mediterranean were exchanged.

In 1946 the shadowy land of Ophir gained some substance. Archaeologists working near Tel Aviv came upon a pottery sherd inscribed in Hebrew script of the 8th century B.C. mentioning "gold of Ophir (belonging to) Beth-horon, 30 shekels." In 1976 after thousands of years of search and speculation it appeared that Ophir had at last been found when a team of United States and Saudi Arabian ecologists announced the discovery of a long abandoned gold mine in an area between Mecca and Medina known as Madh adh Dhahab, or Cradle of Gold.

The scientists who explored the ancient deep mine shafts of this scorched zone examined the thousands of stone hammers and ore-crushing grindstones. Judging from analyses of the huge piles of waste material, approximately a million tons, which has an average gold content of six-tenths of an ounce per ton, the mined ore was richer. The 420 talents brought back from the first voyage to Ophir might well have been in the form of nuggets, crystals and wires easily recovered by panning and winnowing. Geologists speculate there was indeed enough to account for the precious gold to make pots and pans for Solomon's household, as well as for regalia and liturgical objects.

The thousands of stone tools that litter the wilderness of the vast rock plateau along the Red Sea slopes were used later in mining when the surface gold had been exhausted, and might account for the 21 tons of gold brought at a later date. The Madh adh Dhahab lay on the 4,000-year-old north-south caravan route and surface gold would have attracted passersby who brought work to Palestine.

The site fulfills the two chief criteria for authenticity. It is big enough; in fact the mine was reopened from roughly 660 A.D. to 900 A.D. and again from 1939 to 1951 and produced 60 tons of gold during this latter period. And recent study indicates the mine and adjacent areas still contain significant amounts of gold. Several mining companies have applied to the oil-rich Saudi government for permission to exploit them. In addition, the Madh adh Dhahab was within reasonable range of the Palestinian port of Ezion Geber, lying a mere 372 miles south of modern Aqaba and 149 miles inland.

The Book of Job contains a fascinating account of early gold mining which was probably based largely on Phoenician sources and emphasized the amazing length of the hand-hewn subterranean galleries. "There are mines for silver and places where men refine gold; where iron is won from the earth and copper smelted from the ore; and the end of the seam lies in darkness, and it is followed to its farthest limit. Strangers cut the galleries; they are forgotten as they drive forward far from men. While corn is springing from the earth above, what lies beneath is raked over like a fire, and out of its rocks comes lapis lazuli, dusted with flecks of gold. Man sets his hand to the granite rock and lays bare the roots of the mountains; he cuts galleries in the rocks and gems of every kind meet his eye; he damns up the sources of the streams and brings the hidden riches of the earth to light."

Tarshish, another of the vanished gold lands has yet to be found. "The king had a fleet of ships plying to Tarshish with Hiram's men; once every three years this fleet of merchantmen came home, bringing gold and silver, ivory, apes and monkeys," recounts two chronicles. A century after Solomon, Jehosophat, king of Judah, allied himself with king Ahaziah of Israel to build ships at Ezion-Geber for trade with Tarshish. But these ships, built without the help of the skilled Phoenicians, fell apart and could not make the voyage to bring back the gold and beaten silver of Tarshish.

Where was the semi-mythical city of veiled origins? The Greeks gave the name Tartessos to a district of southern Spain where a tribe called the Turti or Turdestani lived. The natives were cultivated Iberians who claimed an ancient civilization and law book dating back to 6000 B.C. The exact site of the city has never been identified but it is thought to have been at the mouth of the Guadalquiver River, farther west than the Phoenician colony of Gades (Cadiz). The first known mention of Tartessos or Tarshish is in the gook of Genesis. In the 10th century B.C. Solomon traded with Tarshish, and judging from references in the book of Ezekiel, the Phoenicians maintained the trade until the 6th century B.C. By 500 B.C. the Tartessan culture probably was destroyed by the Carthaginians who had gained control of the Straits of Gibraltar and were establishing a monopoly of the trade between the Mediterranean lands and southern Spain.

The Iberian Peninsula was one of the greatest sources of mineral wealth in antiquity. "Tarshish was a source of your commerce," writes Ezekiel speaking of Tyre, "ships of Tarshish were the caravans of your imports. There was more than gold from its abundant resources, Tarshish offered silver and iron, tin and lead." The Tartessans figured in the early bronze trade. They appear to have sailed as far as Ireland and Britain, and a number of gold hoards have been found in southern Spain with goldwork that in some respects bears a likeness to early British goldsmithing. In subsequent centuries southern Spain was an entrepot for prodigious amounts of gold, silver, copper, iron, lead and quicksilver —a veritable treasure house which enriched first the Tartessans, then east Phoenicians, Carthaginians, and finally the Roman Empire until Rome fell and mining technology was blotted out in darkness.

As early as 1100 B.C. the Phoenicians sailed 2,300 miles westward to found a colony in the land of the Tartessans. A faded manuscript written in 1580 turned up among some uncataloged papers in the archives of the Indies in Seville not long ago. It gives an account of what the Tyrians found when they landed at Cadiz. The men of Tyre, on a prospecting voyage, landed near Gibraltar where they "mined a vast amount of gold and silver" from the mountains. They sailed away and later returned equipped to found a colony. According to the 16th-century document, while exploring the countryside they came upon a great temple abandoned in ages past by the Tartessans. The men were amazed to find "gold, silver and gems in extreme abundance" in ruined dwellings which appeared to have been abandoned in a panic. "Close to shore," continues the fairy tale, which is typical of gold-filled romances in the 16th century, "the settlers found two great columns of gold and silver which shone like beacons far out to sea."

The author apparently borrowed a bit of his tale from here and a bit of it from there. Inspired by sources which described the actual life-size trees of gold and silver, set with fruits and birds of precious metals and gems, which Persian monarchs had crafted for their courts, he embellished his tale with a wondrous olive tree of solid gold hung with emerald fruit which the Tyrians found in one palatial abandoned house. In perhaps the only truthful statement he concludes "Andalucia was a most important source of both gold and silver for the Phoenicians."

From around 600 B.C., the Carthaginians occupied Cades and exploited the gold mines of southern Spain. Carthage had been founded on the fertile North African coast near modern Tunis in the 9th century B.C. by a band of Tyrians who escaped during a long Assyrian siege of their island city. The colony of Kart-Hadasht or New City, as Carthage was called, grew wealthy and powerful, taking over many of the eastern Phoenician colonies in the western Mediterranean and founding new ones.

They exceded even the Phoenicians in mass production of manufactured goods, although the quality of their goldwork and other crafts was distinctly inferior. They also led the world in wholesale trade, pioneering in cheap goods. Their forte was pawning off gimcrakery for the metals that were a mainstay. Plautus, the Roman playwright, in his *Peonulus*, The Little Carthaginian, pokes fun at the Carthaginians with a comic merchant who sells shoelaces, whistles, nuts and panthers. Roman audiences laughed uproariously at the portrayal of the Punic merchant as huckster and charlatan.

They succeeded in barring the reviving Greeks from sailing through the Pillars of Hercules into the Atlantic, and carved out for themselves an African empire second only to Egypt's in size. Through them the Mediterranean lands received new and generous supplies of African gold, ivory, beasts, slaves and other luxuries. The Carthaginians, trading along the Atlantic coast of Africa for gold, practiced a form of silent barter described by Herodotus. The "dumb commerce" they developed was used in African trade almost to the end of the 19th century A.D. in places where the natives feared slave raids.

"There is a country in Libya," wrote Herodotus, "and a nation, beyond the Pillars of Hercules, which the Carthaginians are wont to visit, where they no sooner arrive but forthwith they unload their wares, and having disposed of them after an orderly fashion along the beach, leave them, and returning aboard their ships, raise a great smoke. The natives, when they see this smoke, come down to the shore, and, laying out to view so much gold as they think the worth of the wares, withdraw to a distance. The Carthaginians come ashore and look. If they think the gold enough, they take it and go their way; but if it does not seem sufficient, they go aboard ship once more and wait patiently. Then the others approach and add to their gold till the Carthaginians are content. Neither party deals unfairly by the other; for they themselves never touch the gold till it comes up to the worth of their goods, nor do the natives carry off the goods till the gold is taken away."

The Carthaginians dealt in North and West Africa gold well into the 2nd century B.C., maintaining complete silence about their activities. Their commercial rivals, the Greeks, continued to believe that beyond the Pillars of Hercules the world came to an end. After the fall of Carthage, Africa became the Dark Continent again, awaiting discovery by explorers yet unborn.

The Carthaginians were not a sympathetic people for all their skills as consummate navigators and genius at international commerce. They may have been much maligned by their bitter enemies the Greeks and Romans, whose references are virtually all we know of them since their libraries were destroyed by Rome. But it is hard to feel drawn to a people who followed the gruesome practice of sacrificing their infant children. Plutarch, the Greek friend of Roman emperors wrote that the Carthaginians "are a hard and sinister people, cowardly in time of danger, terrible when they are victorious...They have no feelings for the pleasures of life."

Both Romans and Greeks shared this view of their common enemy. Greece hated and fought Carthage. Rome finally wiped it from the face of the earth in 146 B.C., sowing salt in the furrows of the plundered city. The Romans gained the treasure chest of Spain, but were unable to continue trading gold on the African coast since Carthage's sea secrets died with her.

Phoenician trading galley under sail, ca. 1500 B.C. Note the bronze ox-shaped ingots on right-hand side.

(Below) Thracian Inlaid gold plaque showing Phoenician influence ca. 5th century B.C.

Author Jenifer Marx, on left, dives for golden treasure on a wreck in the Bahamas where she found a gold coins and a gold scabbard tip.

Phoenician trading ship ca. 1000 B.C.

Phoenician Gold objects—12th century B.C.

Irish gold chalice, ca. 8th century A.D.

Phoenician terracotta figurines of the goddess Tanit, recovered by Bob Marx at Tyre, Lebanon.

(Left) Reliquary of Russian Empress Irene, 1589. *(Above)* Portrait of King Henry VIII in 1542 exemplifying the use of gold, precious stones and pearls on royal vestments, painted by Hans Holbein.

Phoenician gold jewelry; repoussé and filigree. Found in grave in Sidon Lebanon. 5th—4th century B.C.

CHAPTER 7

THE GREEKS AND THEIR CUSTOMERS

Gold, like it or not, has played a leading part in shaping human history. This role greatly expanded when the primeval religious regard for gold was turned into an economic asset by early traders. The perennial desire for gold inspired many of mankind's greatest accomplishments and as many or more transgressions. Gold was one of a handful of materials upon which prehistoric trade was based. The caravaneers who plied the trade arteries through the vast northern forests, across lofty mountain passes and southern deserts achieved far more than their commercial objectives. Along with the river pilots who traveled the watery inland highways and the questing mariners whose fleets coursed the routes of antiquity's seaborne commerce they transmitted the seeds of civilization even as they expanded the horizons of geographical knowledge.

Cultural themes, technology, styles and ideas were spread far and wide in a world without mass media or easy communications. The Eastern Mediterranean had been in contact with the Near East, long gold's center of gravity, for centuries before the curtain fell on the fertile exchange of goods and ideas which nurtured the Aegeans. The Semitic flash of golden brilliance in the era of crumbling empires is set down in gorgeous detail in the Old Testament. The Israelites and the Phoenician sea-kings made the most of the disintegrating Aegean World, the failure of the Hittite Empire in Asia Minor and Egypt's loss of mastery over the Mediterranean. But there were others as well who cherished gold and crafted it superbly during the centuries when Greece was recovering from the disorders and upheavals set in motion by the uncivilized Dorian Greek invaders who had swept south on Mycenae and Crete. And, they gathered the political and spiritual strengths which gave birth to the Hellenistic Period, culminating in the Golden Age of Greece.

In the First Millennium B.C. gold entered the Mediterranean area from sources which previously had furnished chiefly local markets. Foremost among the lovers and users of gold in the emerging Greek world were the Etruscans of central Italy and the nomadic Scythians of the Asian steppes. Both were customers of the Greeks, importing Hellenistic luxuries in exchange for products the Greeks considered vital, not least of which was gold.

The enigmatic Etruscans like the Carthaginians with whom they had close dealings, have come down to us as a somewhat mysterious people. They were established in Tuscany by the 8th century B.C. and by the 5th held sway throughout central and northern Italy from Lombardy to Rome. Their origins are shrouded in speculation and their language, written in Greek characters as yet undecipherable, appears only on a few fragmentary inscriptions. Among the few words known of their language is zamathiman meaning "golden brooch." But in providing for their dead they willed posterity an eloquent inheritance of vibrant funerary art and unparalleled golden artifacts. These treasures recovered from thousands of crypts evoke a vital and sensuous people. Viewing the tomb friezes it is easy to feel drawn to the elegant men and women and the sprightly children who eternally parade in colorful processions, play a variety of games, feast, dance and engage in battle and the hunt. Atop a sarcophagus a sculptor has modeled a couple facing eternity united. The man's arm protectively encircles his wife and both smile slightly. Gold, paint, clay and sculpted stone in the tombs of the nobility bring to life the lives of a paradoxical people whose straightforward love of life and luxury was mingled with an oriental fatalism. The homeland of the Etruscans remains uncertain. The alternatives include Tyre, Troy and

Asia Minor. One theory suggests they evolved entirely from the indigenous Villanovans of northern Italy who moved southward at the turn of the First Millennium. The most plausible explanation comes from Herodotus who wrote when the Etruscans were still on the scene. He stated that they came from Lydia in western Turkey, which later in the 7th-century B.C. grew so immensely wealthy as the middlemen between the trade of the Orient with the Greek colonies of the Aegean coast that the name of Croesus, its last king, became synonymous with colossal riches.

Herodotus wrote that chronic famine in the 8th century forced the populations of this Asia Minor country to divide. While half remained with the king, the rest, carrying their worldly goods, were led by Tyrrenhus, his son, to Smyrna. There they constructed a fleet of ships and sailed west to find a new home. According to the Greek historian they settled on Italy's west coast at Umbria where they subdued the native Villanovans. Mass migrations were not uncommon in the ancient world. The Greeks established trading enclaves on Sicily and the mainland of southern Italy about the time the Etruscan's cities flowered and Rome was traditionally founded by Romulus the wolf-suckled. Those who agree with Herodotus point to the marked similarity of Lydian and Etruscan funerary practices and the oriental mentality of the Etruscans which was at odds with that of the Italian peoples. Oriental influence on Etruscan goldwork is very prominent but can he traced less to Lydia than to imports from Carthaginian and Greek workshops.

Etruscan goldsmiths were among the greatest the world has ever known. The Greeks and Romans who scorned the pleasure-loving Etruscans as pirates and debauched sensualists envied their skill as goldsmiths, metallurgists, warriors and traders. As goldsmiths they crafted some of the most extravagant and original golden fantasies ever made. As metallurgists they mined and smelted utilitarian metals, especially pig iron, which, as traders, they exported to Carthage, Egypt and the Eastern Mediterranean and exchanged for the countless Greek vases found in Etruscan tombs. Their contemporaries considered the Etruscan love of music, evident in hundreds of murals, decadent and excessive. According to Aristotle they kneaded dough, beat their slaves and even fought battles to the sound of a flute. Nevertheless they were talented at warfare. As formidable warriors the Etruscans spread their culture throughout the heart of the Italian peninsula.

The virtuosity of Etruria's goldsmiths shines in pieces which range from the massive to the miniscule. The Etruscan penchant for conspicuous consumption coupled with great affluence based on industry and trade accounted for the lavish amounts of opulent jewelry and goldwares entombed with their dead. Etruscan tombs differed in one important aspect from almost all other gold-using cultures. They contain no ceremonial weapons, although in 1823 a man looking for stones to mend a road discovered a tomb which contained a warrior in full armor with a golden crown fashioned of sheets of fine gold fastened to bronze backing and sculpted with lilies.

The high status of women in Etruria is evident in the volume and opulence of the gold ornaments and furnishings with which they were buried. Wealthy matrons, wives of merchants and nobles enjoyed status and influence far greater than the modest women of Greece or early Rome. They banqueted with the men, entertained guests, proposed toasts and drank in public. The austere Greeks and severe Romans naturally were scandalized by what they considered lascivious and shocking behavior. In Greece women were kept in virtual seclusion, and a Roman was acquitted for the murder of his wife provoked by seeing her taste wine. The flamboyancy of the sybaritic Etruscan females gave rise to the belief that maidens raised money for their dowries through prostitution, and the Romans wrote that the shameless Etruscans made love in public.

Most characteristic of the Etruscan goldsmithing were the safety pins called fibulae with which both men and women fastened their garments. They varied in size and style from the sublime to the decidedly vulgar, changing with fashion. In some periods they were rather simple by the Etruscan standard which valued complexity, relying on the intrinsic loveliness of the metal. More often they were lavishly decorated with the most gossamer filigree, repoussé or the granulation at which no one has ever excelled the Etruscans. Sometimes they were ridiculously large or unforgivably over-ornamented by goldsmiths who obviously loved what they were doing and didn't know how to stop. One amazing fibulae found in the tomb of a noble lady was over a foot high. It was part of a spectacular treasure in gold, ivory and amber that included long earrings formed of several pieces and elaborately covered with patterns in repoussé. Her bones had crumbled but not the rings, bracelets and many-pierced gold plaques, each exquisitely crafted, which may have once covered her entire robe. The most splendid ornament in her grave was a gold breastplate, 16 inches high completely covered with typical abstract designs interspersed with foliage and flowers.

Goldsmiths created dazzling three-dimensional compositions to decorate the fibulae the Etruscans so adored. On one, a goldsmith wrought an almost miraculous procession of meticulously modeled horses, lions and sphinxes in the round. They are embellished with granulation and parade up the bar of the pin. Another fibula, also covered with animals, ends in a curved fastening which is a base for five ferocious lions held in check by two ornate restraining bands.

The nature of their imports is reflected in the overwhelming amount of Greek artifacts found in such tombs as Praeneste, Caere and Tarquinia. They were so enamored of anything Greek that Etruscan tombs of the wealthiest era (600 to 400 B.C.) have proved a more fertile source for Greek jewelry, statuary, ceramics and ornaments than Greece itself. Shiploads of Greek merchandise were unloaded at the Etruscan ports to satisfy a ravenous appetite for imported luxuries. Some of the gold jewelry and many of the gold and silver-gilt bowls excavated from Tuscany and Latium are chased or otherwise decorated with scenes of Greek processions or Egyptian battles. Some of the pieces are imported and others are native copies. The Etruscans admired almost everything foreign and incorporated Assyrian and Greek geometric patterns, Egyptian lotus-flower bases and various Anatolian and Cypriote elements as well into their own gold work.

Native Etruscan motifs appear as geometric patterns on gold plaques found in early Etruscan tombs. But from the outset the more sumptuous pieces of goldwork incorporated oriental elements from a wide variety of sources. Some 8th-century B.C. tombs contain Egyptian scarabs. By the 7th-century, luxuries were imported and then copied from by the Etruscans. During this same century Phoenician artisans began working in Etruria crafting bowls and jewelry in gold and silver, during what is called the Orientalizing Period. Later, gold alone was preferred, although silver was locally mined in Popoulonia and used there and elsewhere for the earliest issues of coinage. Imports from Greece, Cyprus and Phoenicia were not the only ones to influence the arts of the Etruscans. Lion motifs are traceable to the region of southeast Anatolia where the Hittites lived and the processions of winged bulls, stags and horses found on hammered bronze work as well as on pieces of precious metal had their origin in the art of Crete.

Another impulse on the 7th-century B.C. came from the kingdom of Urartu in what is now Armenia and Eastern Turkey. It is thought the metalwork of this area reached Etruria via Ionia and the Black Sea colonies of the Greeks, the same way it flowed to the temples of Archaic Greece.

Ivory was a material introduced during the Orientalizing Period that found favor with the Etruscans. Hippo and elephant ivory from Phoenicia or Carthage was carved locally into small caryatid figures holding braids, combs, plaques and the bucket-shaped vessels called situlae, which were more commonly of pottery or bronze. Amber was known much earlier. Beads and parts of amber fibulae have been found in Iron Age tombs. Villanovan graves of the early 7th-century contain tiny carved monkeys and women-shaped amulets of amber. In the late 6th and early 5th centuries amber was in fashion. People who believed in the magical and curative properties of the vegetable gold wore amulets mounted on fibulae. The natural shape of pieces of translucent amber were taken into account in carving magnificent miniature groups of reclining people, some of which have been found in the region of Ancona where they may have been made.

Goldsmiths fashioned rings and earrings paved with colorful semi-precious stones and intricate gold disks for attachment to clothing. They made bracelets, embossed beads, pendants, exquisite chains and necklaces. Examples using granulation and filigree wire as well as plaited fine gold wire of large-scale pieces include a gold covered coffer from Vetulonia where 15 early well-tombs were sunk in circular formations. They were found to be overwhelmingly rich in gold, silver and bronze objects and the earliest exotic products. Later tombs yielded large openwork coverings of gold animals that covered wooden situlae. Pieces of jewelry were also made by sprinkling golden blossoms and leaves on an intricate net of gold wire.

The Etruscan goldsmith could inlay gold with enamels or beat out a perfectly plain gold cup with decorated handles, which was in great contrast to the exceedingly opulent fibulae overloaded with decoration. The genius of the Etruscan smith was most apparent in the performance of granulation. Miniscule globules of gold, as small as 1/1,000th of an inch in diameter so that as many as 180 fitted on an inch long line were affixed in the characteristic geometric patterns or in clusters, cones or lines on the gold surface of jewelry and vessels. London's Victoria and Albert Museum displays a golden bowl from the late 7th-century B.C. found at Praeneste which has an intricate and dense pattern in granulation. The breathtaking effect has been likened to that of light caught on early morning dew. The approximately 137,000 miniature spheres were applied to the bowl by a process lost until an Englishman in 1933 accidentally rediscovered it when the fish-glue he was boiling in a copper pot boiled dry. H.A.P. Littledale saw that copper oxide had formed on the inside of the pot and when it was reheated the fish-glue absorbed the copper oxide. The Etruscans and others who were ancient masters of granulation glued the gold balls into place one by one using a copper-salt mixture. When it was heated, the glue disappeared leaving a microscopic bond of copper between the gold sphere and the gold base.

The Etruscan trade with Carthage brought both Spanish gold and gold dust from the African trade to the local workshops. But a good supply of alluvial gold was also from closer at hand from the upper part of the River Po whose tributaries run down from the western Alps. Central France had large amounts of alluvial gold which could easily be panned and it is thought some of this Gaulish gold reached the Etruscans as well.

In a short span of time the highly developed Etruscan culture of traders and warriors came to fruition. Although it boasted large, well-appointed urban centers, Etruria was never a unified nation. Like Phoenicia it was a loose confederation of about a dozen regional capitals which shared with a total of some 50 cities whose wealthy citizens were princes, priests, merchants, brokers, shippers, traders and goldsmiths. They all shared a common religion, language and culture and fought together against their common enemies. Etruscan wealth was founded on farming and sustained by piracy and, above all, international trade in metals. Their land was not only fertile but laced with gold, silver, iron, copper, lead and tin. The Etruscans exported fine bronze weapons, cauldrons, pails, cups and less useful but equally coveted pieces such as figurines, candelabra and the wreaths of ivy and laurel leaves worn in the lively Etruscan banquet scenes shown in the tomb murals.

But the heart of Etruscan metallurgy was in the iron mines of the Central Italian mainland and on the islands of Elba and Sardinia. Iron became essential to the kingdoms of the Mediterranean perpetually in conflict with one another. A great deal of iron came from Elba where the ore lay near the surface and was so abundant that the Etruscans could be excused for thinking that it magically replenished itself. At first the ore was smelted on the island but later a processing plant was put into operation at Populonia, a bustling harbor on the mainland coast, 80 miles from Telamon, the naval station from which the Etruscans controlled the Western

Mediterranean trade routes. Huge mountains of slag accumulated here. After World War I when Europe was starved for iron, these towering slag piles were profitably worked for iron left behind by the Etruscan process.

In addition to huge amounts of pig iron the Etruscans traded raw copper, fine bronze articles and the shiny black pottery called bucchero ware with the Greeks, Greek colonies, cities in the Eastern Mediterranean and with the Carthaginian ports in North Africa, Cyprus and Spain. Control of the west coast of Italy and the Western Mediterranean sea lanes was fought over by the Greeks and Etruscans, the Greeks and Carthaginians,s and the Carthaginians and Etruscans. But by the 7th and 6th centuries B.C. Etruria was losing control of the seas to the Carthaginians and her role as an international force declined, she turned to internal conquests within Italy. Etruria's domination of the Western Mediterranean trading lanes, so vital in the Iron Age, eventually crumbled before the superior Punic navy, but the two powers maintained amicable relations for some time. Carthaginian vessels carried Etruscan merchandise and brought gold and luxury items hack to Italy. With Etruscan help the Carthaginians were able to establish a firm toehold on Corsica and partial control of Sardinia thus the routes west to Spain's mineral wealth were shut off to the Greeks.

Proof of Punic-Etruscan relations came to light in 1964 when one of the most fascinating of recent archaeological discoveries was made at Pyrgi, the port used by the inland Etruscan cities. One of the chief cities was Caere, famed for its gold work, which once had 25,000 inhabitants. It was reputed to be one of the richest cities in the entire world at that time. Classical authors made frequent references to a temple on the shore at Pyrgi which was bedecked with gold and had a great treasury filled with gold and silver. Archaeologists discovered the temple and among the artifacts they found three folded gold sheets. Two of the beaten gold plaques bore Etruscan inscriptions and the third was in Punic. They were written around 500 B.C. and in parallel texts record the dedication of the temple by the ruler of Caere to a goddess worshipped by both peoples. She was Uni to the Etruscans and Astarte to the Carthaginians, the goddess of fertility and sexual love. The golden texts indicate that there were some Carthaginians living at Pyrgi. Twenty-nine tiny nail holes pierced the gold sheets, which may have been removed from the sanctuary during an attack and buried for safekeeping. Whoever took them down carefully placed 29 diminutive nails in the folds of the eternal letters.

As long as Etruria dominated the shipping routes, she could charge what the traffic in strategic materials would bear. The wealth poured in to feed the insatiable demands for conspicuous consumption of luxuries of a people their contemporaries called crass and materialistic. As profits mounted, gold and treasures accumulated and as their owners died they were buried with them.

In the European Dark Ages, which followed the fall of mighty Rome in the 4th-century A.D., bands of hungry outlaws plundered thousands of the more obvious Etruscan tombs, making off with gold and jewelry, smashing pots and other artifacts deemed of no value. By that time the Etruscans had been assimilated into the Roman World. Rome's first historical character was an Etruscan, Tarquin the Proud. For a century following Tarquin, Etruscan kings ruled the nascent city. Around the beginning of the 6th-century B.C. they were driven out but left a strong imprint that colored Roman life for centuries. Etruria bequeathed Rome her religious practices, engineering projects of roads and flood control, dress and facets of social organization. The Tarquinians laid out the city's sewer system, built bridges and temples. They drained the swamps between the Capitoline and Palatine Hills and created the Forum where the courts, markets, senate and citizens' assembly governed the life of the crude city that grew to empire.

Unfortunately, in spite of the wealth, power and natural advantages the Etruscan cities had in common, they were unable to suppress internal squabbling long enough to confront together the growing might of Rome. Relying on mercenaries to do battle for them, they lost crucial encounters and by 280 B.C. the Roman Republic had crushed every one of the cities of Etruria.

Etruria's sun set with the blazing dawn of Roman power. For several centuries the Etruscan language survived as a rustic dialect in the Tuscan hills, but imperceptibly every trace of the culture that gave Italy its first taste of grandeur vanished. In the Renaissance there was a flurry of interest in the shadowy Etruscans, but it was the fantastic discoveries of the 18th and 19th centuries, coming from the long lost burials in the malarial Tuscan swamps, which made a jaded world take notice. For countless centuries Etruscan tombs had been robbed of their golden treasures in a random manner. However, in the 1820's the greatest treasure hunt of the century was triggered by a peasant who was plowing a field about 40 miles north of Rome at Vulci. He crashed through the roof of a buried necropolis with his team of oxen and landed on top of Etruscan gold.

The property containing the necropolis had been sold by the Pope a few years before to Napoleon Bonaparte's brother, Lucien, Prince of Canino. The prince, with absolutely no regard for aesthetics or history, began to ruthlessly cannibalize the Vulci necropolis which set a precedent for other landowners. He worked at it thoroughly as if it were a gold mine. Indeed, it was a gold mine. In four months of indiscriminate looting his laborers recovered more than 20,000 relics of gold and silver, more than any other ancient site in the world had ever produced. Soon anyone with a plot of ground in the heart of Italy was scrambling to search for buried gold. It was easy work; finding the telltale clues which signaled the presence of a tomb, breaking in and then carrying out the loot which dealers in Rome and Florence fell over each other to buy. The antiquarians bought cheap and sold dear to avid collectors from all over Europe and even the United States.

Two men stand out from the greedy throng. George Dennis, the founder of British Etruscology, was consul in Italy in the 1840's and one of the few men interested in the Etruscans as a people. He devoted his life to piecing together their history as it was revealed through the art and artifacts of the tombs. The other savant was the immensely wealthy English numismatist, Richard Payne Knight who died in 1824. His collection of gold, jewels, coins and ancient bronzes forms the core of the British Museum's great Etruscan collection. Dennis was present at the opening of a burial site at Vulci in 1843. He watched the Prince's workmen as they labored under the vigilant eye of an armed superintendent. The men found dozens of whole bucchero vessels, with their lovely metallic sheen intact. They and all other pottery were smashed on orders from the Prince who wanted only the precious artifacts removed. The overseer made sure that the pottery, furniture; the entire contents of the tomb were destroyed. Dennis couldn't even persuade him to spare one vessel as a souvenir.

This ghastly act of desecration was repeated at each site. In spite of the hundreds of thousands of artifacts that were smashed, thousands managed to survive. The Vulci tombs were worked for 150 years and an estimated 85 percent of the Etruscan vases in Florence and Rome came from the 6,000 or so tombs there. The rage for things Etruscan spawned an industry in turning out fake gold jewelry and replica ceramics. There is still a black market in Etruscan objects d'art and an even brisker trade in counterfeits made and aged by talented Tuscan forgers, descended perhaps from the distinguished craftsmen of Etruria.

Nineteenth-century ladies had a passion for the lovely, showy gold jewelry that once adorned Etruscan matrons. Lucian Bonaparte's wife inaugurated the vogue by wearing an incredibly beautiful assemblage of pendants, jeweled earrings and armlets which made her the envy of society women. The products of the finest ateliers of Paris or Vienna couldn't hold a candle to the exquisite golden ornaments that had glorified lusty Etruscan beauties in life and comforted them as they lay forgotten through the ages under the black Italian earth.

While the Etruscan cities were still young, another trade connection the Greeks were crafting prodigious amounts of gold. Far from the mild Italian peninsula the horse-riding Scythians, wanderers of the vast Eurasian steppes came into contact with the Greeks at the trading colonies Greece had established on the shores of the

Black Sea. The Scythians, who came from Central Asia in the 8th century B.C., founded an empire based on a federation of separate tribes sharing a common culture. They exchanged grain, vital to feed the population of infertile Greece, for Greek goldwares and other Attic luxuries which have been found in the barrow burials of warrior chieftains west of the Volga and north of the Black Sea.

The Scythians are best known today for their art—the rhythmic, bold animal style of the grazing lands which was translated into golden expression of great magnificence and abundance. Scythian art impulses spread westward with successive waves of migration to shape the developing Celtic art of Europe. They spread to the Southeast as well with nomad immigrants to Luristan, the region of the central Zagros Range between Iraq and Iran.

In antiquity, however, the Scythians were famous not so much for their dazzling golden displays as for their savage ferocity as warriors. The Old Testament, once again, records a chapter of ancient history with clarity and passion. The Book of Ezekiel prophecies God's triumph over the World—by which he meant Israel's victory over her enemies. Its author recalls the time in the early 7th century B.C. when Scythian forces whirled down on the Levant, wasting and scarring Syria and Judea and halting at Egypt's border only when they were bought off with massive amounts of Nile gold. Speaking to Gog, the name by which he calls the Scythians, Ezekiel foresees a time when "....with many nations from your home in the far recesses of the north, all riding on horses, a great host, a mighty army, you will expect to come plundering, spoiling and stripping bare the ruins where men now live again, a people gathered out of the nations, a people acquiring cattle and goods, and making their home at the very center of the world. Sheba and Dedan, the traders of the Tarshish and her leading merchants will say to you, 'Is it for plunder that you have come ? Have you gathered your host to get spoil, to carry off silver and gold, to seize cattle and goods, to collect rich spoil?'"

The most detailed information about the Scythians and related tribesmen comes from that remarkable prolific source Herodotus, who, in the 5th century, visited the city of Olbia, the "Paris" of the Black Sea. Olbia was a very wealthy mercantile port at the confluence of the Bug and Ingul Rivers, founded by Miletus in the mid-7th century, which shared in the lucrative commercial traffic between Asia Minor and southern Asia. The Scythians figured prominently in the history, not only of Syria, Judea, Egypt, and Greece, but also of Assyria and especially of Persia. They had denied Darius the Great a crucial Persian victory in his bid to conquer Greece. Herodotus went to Olbia to collect data for his "History of the Persian Wars" and ended up with a lengthy ethnographic treatise on the religion and customs of the nomads, which he incorporated into the fourth volume of his history. Much of the information he must have gained second-hand from the Ionian Greeks traders with the barbarous horsemen who supplied the bulk of the wheat and cattle for Greece.

By the nomads' own account they descended from the son of Zeus and a creature, half snake and half woman, who was the daughter of the Dnieper River. This ancestor had three sons who ruled the land and during their reign four implements, all of gold, fell from the sky—a plough, a yoke, a battle axe and a drinking cup. Each brother in turn tried to pick up an implement, but at the approach of the elder two the gold blazed up. The youngest brother was able to pick them up without the flames breaking out, and thus he became the paramount ruler and forefather of the Scythians.

The Scythians' history is one of migration in response to pressure starting in the Asian heartland. In the 9th century they and kindred tribes were evidently concentrated east of the gold-rich Atlai Mountains of Siberia. In the 8th century the emperor of China sent a force to push back a savage horde called the Hsiung-nu. As they retreated westward they dislodged the Massagetae from their traditional grazing lands north of the Oxus River, an important source of ancient gold. The retreating Massagetae spilled into the Scythian area, sparking a massive migration westward.

The Scythians were among the world's first and best horsemen.

They spread across the Oxus with terrifying swiftness, pushing the Cimmerian foot soldiers out of the Caucasus and the plain north of the Black Sea where they had led a prosperous existence for many generations. Their fame spread with their conquests. They gave no quarter, forcing the Cimmerians to retreat for more than 30 years until they were wiped out in Lydia. The Scythians dominated most of Asia Minor during the 7th century B.C. from a capital between the Euphrates and the Caspian – Saqqez – from which they boiled down to make the raids mentioned in the Bible. They pillaged, burned and exacted tribute without mercy. Herodotus wrote, "Their insolence and oppression spread ruin on every side."

The Scythian chieftain Partatua formed a temporary alliance with Esharhaddon, emperor of Assyria, and married his daughter. In 1947 a gold hoard of great beauty and value was found on a mountain towering over a remote village in northwest Iran. Some scholars think part of it may be a portion of the dowry of the Assyrian princess. It was discovered by chance and almost lost through the gold-greed of Iranian villagers. A shepherd boy, walking near the town of Ziwiye, stopped to investigate something shining among the rain-washed ruins of an ancient citadel. As he was grubbing in the earth, extracting what proved to be several pieces of worked gold, a merchant passed by and saw him. He convinced the ingenuous boy to part with his gold for a few coins. The man buried the gold in his yard while planning what to do with it. However, the lad had told his parents of his discovery and the subsequent transaction. They went to the village chief with the story. The headman led a group of men to the merchant's house where they beat him and dug up the gold. Then the villagers went to the site and unearthed many more pieces of gold. By the time the authorities leearned of the treasure the peasants, ignorant of the archaeological and artistic value of the gold, had broken up the pieces and divided them. With great difficulty the fragments were assembled and given to experts who spent many months trying to restore them. Eventually, and almost miraculously, the collection was put on display.

The gold work included a funnel with a curved handle, many gold bracelets ending in animal heads, gold furniture terminals in the form of beasts and birds and a gold hand net with two finger rings. The gold reflects influences of the Assyrians, Scythians, Mannians, Urartians and even Phoenicians. A number of repoussé and chased gold pectorals were found, one of which is a large ceremonial breastplate with two rows of monsters and animals flanking the Tree of Life. The fantastic creatures are a mélange of Assyrian and Phoenician styles, whereas some of the animals are distinctly Scythian in feeling. The villagers destroyed beyond repair a gold plaque almost a foot long and over six inches high which weighed almost a pound. It is now reduced to 23 fragments decorated with reclining mountain goats and stags in repoussé. What was once a magnificent 4½-pound silver bowl covered with gold appliqués of Scythian inspiration was also split among the treasure hunters. All that remains is a fragment with gold lynx-like creatures, flower petals, birds of prey and geometric forms.

The Ziwiye trove, as it is known, contained many gold and ivory objects representing a blend of various styles including Phoenician, Assyrian and Scythian of the 8th and 7th centuries B.C. and was more likely buried by a high Assyrian officer. It also could have come from the Medes, masters of Persia whose king Cyaxares, after 600 B.C., forced them to relinquish their hold on most of Asia Minor. Some of the Scythians withdrew to settle in southern Russia where they controlled the area between the Persian border and the Kuban—an area between the Caspian and Black Seas. Some settled between the Caspian and Aral Seas where they intermarried and gave rise to the people known as the Parthians three centuries later. A third group infiltrated India and founded kingdoms there.

The Scythians in the Kuban area and southern Russia prospered, but they never entirely gave up their nomadic ways. In spite of continuing to raid and loot caravans, they became agricultural magnates, exchanging livestock and grain at the Ionian cities on the Black Sea for weapons, gold ornaments of Ionian manufacture and the famed sweet Olbian wine. They grew very rich and consequently buried their dead with much gold. Graves from the 7th and 6th centuries B.C. excavated in the Kuban were filled with the treasures belonging to the wealthy, powerful chieftains and families. In the 6th and 5th centuries the richest burials were in the Crimea and southern Russia and belonged to the group known as the Royal Scyths who ruled the area. They wielded economic power until the first century before Christ, managing to exact annual tribute from some of the Pontic Greek cities in exchange for not invading them. In the 4th century a new shock wave of horse-riding tribesmen from Central Asia appeared on the Don River. These Samarians, whose women fought alongside their men, occupied much of southern Russia and even penetrated beyond the eastern Balkans.

They too were great lovers of gold and featured skilled smiths. Hundreds of their tombs abounding in gold were found in the 19th and 20th centuries. They yielded horse trappings, arms, gold plaques for trimming clothes which were embossed with geometric designs typical of Samaritan art, fibulae, little gold bottles covered with jewels set with colored glass inlays, gold cups and vessels—some of them signed in Greek by Samaritan jewelers—of an eloquence and refinement which belies their reputation as barbarians. Tombs and buried hoards contained geometric, floral and animal designs, decorated diadems, bracelets, pendants, necklaces, buckles, plaques, torques, buttons and rings. Samaritan goldsmiths were at ease with a number of demanding techniques, including insetting stones, inlaying glass, repoussé and even gold wire filigree. A second burial, so rich it was probably that of a queen, was discovered in 1864 at Novocherkassk. The skeleton was adorned with a magnificent crown, gold bracelets and necklace. In addition, there were gold and silver cups of Samaritan manufacture and several signed pieces from Greece showing the links between the two people. One very elaborate gold bottle featured an all-over repoussé design of lions, elks and monsters. The most charming object was a small agate bottle in the form of a lion. The body is a polished cylinder of agate and the head, forequarters and hindquarters are modeled in gold.

The Samaritans crossed into Scythian territory at a time when the Scyth king, a 90-year-old warrior, had died in battle against Philip II of Macedonia in 339 B.C. They kept relentlessly at the Scythians until by the 2nd century they had confined them to the Crimean area, and before the century was out had succeeded in supplanting them entirely, wiping out the last vestiges of the once mighty Scythians.

The Scythians were of such consequence that they had been able to repel a Persian invasion in 513 B.C. and force Darius the Great into an undignified, hasty retreat. Darius planned to starve Greece into submission, which meant gaining control of the Scythian territory, the chief agricultural supplier for Greece. Darius was so confident of the outcome that he led a large Persian army across the Danube to drive the horsemen from the steppes and sever the supply line to the Greeks that he boasted he would return victorious to the Danube in 60 days. Instead, he found no way to catch the fleet horsemen in their Greek chain mail and colorful felt garments flashing with battle gold. The nomads scorched the earth and poisoned the wells as they moved like quicksilver, always just out of reach. Finally Darius, weakened by slashing enemy attacks on the vulnerable Persian flanks withdrew to save his army.

It was this crucial Persian defeat which brought Herodotus to Olbia to gather material for his book. "The Scythians," he wrote "use neither iron nor silver, having none in their country; but they have bronze and gold in abundance." Actually they did use some iron along with silver but gold was by far the most significant metal to them. "The Royal Scythians guard their sacred gold with most especial care, and year by year offer sacrifices in its honor." The Scythes and related tribes were overwhelmingly rich in the most precious and durable of metals and they used it imaginatively in many ways.

Some of the immense amounts of gold available to the virtuoso smiths of the steppe dwellers came from the rivers of Southern

Russia. Some probably reached them from the Oxus Valley; perhaps they even received gold from the Kyzyl Kum desert of Uzbekestan, where the Soviet Union today processes millions of tons of auriferous sands each year using water piped 700 miles from the Oxus (Amu Darya) River. Most of the gold, crafted into combs for the long hair of warriors and death masks for their horses, came from the incredibly rich Ural-Altai area, the wild, desolate region in the central part of Asiatic Russia which enriched the Czars who exploited it with serf labor in the early 19th century A.D. Even now the mountains, plains and deserts of this region yield copious amounts of high quality alluvial gold. To the east comes gold from the Kolar region of Siberia where thousands of forced laborers perished in the bitter cold producing gold for Stalin. Although Soviets are very secretive about the annual output of their gold operations, Western economist believe it is as much as 250 metric tons. This would make the Soviet Union the worlds second largest producer, although annual production is little more than a quarter of South Africa's. These zones were worked in the most ancient of times. An 18th-century traveler in Siberia wrote of visiting ancient gold mines and of finding evidence of the miners who had worked with copper and stone tools, wriggling through the tight tunnels to get at the quartz from which they used sharpened boar fangs to extract filaments of gold.

The literature of Classical antiquity abounds with horrifying tales of grotesque monsters and magical beasts that inhabited remote zones of the earth. They are described as pygmies, headless people, men whose ears hung to their knees and so on, and their habitats were pinpointed on maps into the Middle Ages. A fair number of these mythical beings figure in the history of gold, particularly Scythian gold. Most famous of all were the one-eyed Arimaspi, a word of Scythian origin, who were first mentioned before Herodotus in a Greek poem. Without believing it himself, the historian described the folk who dwelt in the far north where feathers (Snow) filled the air eight months of the year. They were horsemen allegedly having only one eye (probably a forehead oil lamp of ancient Siberian miners) who lived near a river filled with gold.

The gold was guarded by especially ferocious griffins. These winged monsters who were a favorite with Persian goldsmiths were often shown with the body of a lion, two pairs of powerful wings and the head and shoulders of a giant bird of prey. Aeschylus called them "the hounds of Zeus who never bark, with beaks like birds." It was believed they attacked whoever went after their gold, shredding the thief with their razor-like beaks or carrying them aloft in their relentless claws to drop them to their death on the jagged mountain peaks below. "Grypos" is the Greek word for "Beaky" and some have linked the griffins to the beaky-nosed Mycenaeans and their Achaean kin who may have penetrated Western Asia looking for metals before they rose to prominence in Greece. Another explanation suggests the griffins represented a vestigial minority of prehistoric beasts of the Ice Age. It is more likely however that the Scythians wished to protect their gold sources and invented the blood curdling tales to keep competition at bay.

In life the Scythian warrior horsemen ostentatiously sported all manner of gold ornament and trappings. So did their beloved horses, the very core of their culture. The nobles had vast herds of horses or ponies but every Scyth had at least one fine mount on which he lavished much attention. Saddles covered with brightly dyed felt were often embellished with gold plaques. Cheek plates and bridle frontlets were often fashioned of gold in open-work animal shapes, and reins, bridles and other trappings were covered with embossed gold. The horses that were buried with a warrior of high rank were richly caparisoned in gold. In the grave of one Scythian chieftain who was buried with golden appliqués on his costume and masses of gold objects for the afterlife were found the remains of 360 horses.

Lustrous gold was also used for jewelry, vessels, arms, shields, helmets, belts and girdles. When a Scyth died he was entombed with the comforting presence of his favorite concubines, horses and masses of beautifully worked gold objects in two very distinct styles—the native animal style and the goldwork commissioned by the nomads from the refined Greek smiths working in the Black Sea cities.

Herodotus told the world about the Scythian hemp-inhaling ceremonies, the quivers they made of human skin and many other fascinating aspects of their lives. In describing the burial practices of the Issedonians he reported that when a man's father died, his body was chopped into pieces and mixed with the flesh of sacrificial sheep and partaken of by all of the deceased's relations at a great banquet. His head was "stripped bare, cleaned and set in gold," turning into a prized relic which was brought out for an annual celebration.

He wrote that when a king died among the Scythians, he was embalmed and his abdominal cavity stuffed with a preparation of herbs, after which the body was enclosed in wax and carried about on a wagon for all the tribesmen to see. Warriors who viewed the dead king mutilated themselves, cutting off a bit of ear or piercing themselves with a blade or arrow. When the circuit of the tribes was completed, the king was laid in a deep tomb. His favorite concubine was strangled and laid with him, "and also his cup-bearer, his cook, his groom, his lackey, his messenger, some of his horses, firstlings of all his other possessions and some golden cups." Sometimes in a ritual a year after a chieftain's death as many as 50 men and 50 horses were killed and stuffed with straw; the riders impaled on their mounts which were suspended on poles between stakes arranged in a circle around the burial mound and left to keep macabre watch over the dead prince.

Even though the Scythians' used gold until 1715—when a most unusual christening gift was given to the infant heir to the Russian crown—no one had a glimmer of their profusion of gold or its rare and startling beauty. A nobleman presented the son of Peter the Great with 20 ancient gold plaques, some cast gold and some embossed with repoussé designs, all of which had been discovered in kurgans—the burial mounds of Siberia. The emperor was fascinated with the plaques shaped into animals. The beasts, which may have been shield ornaments, impressed all who saw them with their unusual loveliness and great vitality. Along with later gifts of animal-style gold objects the plaques were put in a museum established by the Czar who forbade unauthorized excavation of the kurgan mounds dotting much of the landscape and the melting down of ancient gold artifacts.

Since that time literally thousands of splendid pieces of nomad gold have been found ranging from the frozen Pazyryk burials in the Altai Mountains near the border between China and the Soviet Union to the nation's western border with Europe. The greatest concentration of burials can be found in and around the Crimean Peninsula of south Russia. For over two thousand years men discredited the tales of the passionate riders who drank from the gold-lined skulls of their enemies, made cloaks of human skin and inhaled the narcotic fumes of burning hemp seeds which made them "shout for joy." Archaeological evidence from the steppe kurgans corroborates Herodotus' arresting accounts which would otherwise seem quite incredible.

What is incredible is the remarkable manner in which the love of gold by these so called barbarians was translated into objects of such loveliness and intensity that they have the power even in the 20th century to evoke a strong feeling for a lost age of beauty and ferocity. Scythian gold falls into two general categories; the indigenous style known as Animal Art of the Migratory Peoples and the extraordinary Hellenic pieces executed by Greek goldsmiths in Asia Minor. The Asiatic horsemen had keen appreciation for the subtlety of Greek art and prized the pieces reflecting nomadic themes made by the goldsmiths of the Black Sea settlements. The Animal style with its compelling force and beauty grew out of the prehistoric wood and bone carvings made by nomads ranginge from China to Persia. The Scythians seldom depicted human figures or even the cattle and horses upon which their free-wheeling life style hinged. They preferred images of stags, which had been a religious totem in pre-Scythian Siberia, griffins, goats and other real or imaginary creatures. The mythical beasts, often blended with elements of real creatures, were of ancient origin but the Scythes interpreted them in an unusual fashion. A strong element of sympathetic magic may have accounted for this. Perhaps they felt that a gold image of a swift, strong or fierce creature on a warrior's weapon, mount, armor or person imparted invincible qualities. Their familiarity with the animal kingdom allowed the steppe goldsmiths to infuse their creations with a strong feeling of movement and energy, often compressed into a closed object with an astonishing instinct for form.

Violence was an important element in the life of the Scythians and dominates their art. Whoever gave quarter on the harsh steppe lands was lost. Gold animals are characteristically shown locked in mortal combat. Themes such as griffins attacking mountain goats, which emphasize the helplessness and vulnerability of the attacked and the power of the attacker, are common in the art of the Near Eastern civilizations and the Scythians may have been influenced by Achaemenid Persia which lay on the western border of their grazing lands. Recumbent or battling stags and felines were a favorite with Scythian goldsmiths, particularly for decorating weapons or the round shields many Scythian fighters used. Most often they were depicted with the legs contracted perhaps to indicate that the creature was dead or passive. Another convention was the incorporation of smaller creatures within the silhouette of a larger animal as shown on a gold plaque from Peter the Great's collection. The 6½-inch piece depicts a tiger locked in a deadly tangle with a horned predatory beast whose mane and tail tip are formed of five miniature stylized griffin's heads. Another example is a somewhat abstracted recumbent stag, 12½ inches long, which may have been a shield emblem. The contour of the rear haunch becomes a griffin's head, and along the beveled planes of the stag's unnaturally long throat a lean greyhound stretches, and a dog, a lion and a hare are worked in repoussé inside the body outline. The antlers have nine tines, a magical number, forming a serrated pattern, and behind the last of them is a ram's head with down-curving horns. This piece was commissioned from a 4th century B.C.

Greek smith and was found in a tomb in the Crimea and shows that Greeks did Scythian themes on order.

Along the Black Sea the Greeks, for the most part, enjoyed a peaceful co-existence with their Scythian overlords. The nomads with whom they had so little in common admired their skill at working gold and the Greek goldsmiths took pains with sumptuous pieces based on Scythian iconography which also reflected Hellenistic, Persian and Armenian (Urartians) influences. The Scythians didn't abandon the Animal style with its visceral magic appeal in favor of the refined Hellenistic work. The two co-existed and animal totems and exquisite Greek pieces of breathtaking delicacy were often buried in the same kurgan.

Far and away the single most stunning piece of Scythian gold was a 12-inch wide gold pectoral discovered in 1971. It is of 4th century B.C. Greek workmanship and is a masterpiece in miniature, perfectly proportioned and blending the measure and delicacy of Hellenism with the realism and rolling life of the mysterious hostile steppes. It contained 48 individually cast figures ranging from humans to grasshoppers; the artist had created a virtuoso contrast of Scythian domestic life with the harsh wilderness outside the encampments. The exquisitely crafted figures are soldered onto the top and bottom of three registers separated by four rope coils of diminishing diameter, tapering to the curved ends of the pectoral which are hinged and elaborately worked with gold mesh, repoussé, and chasing. The center zone is filled with sheet gold and richly decorated with flowers, buds, leaves and tendrils on which four sculpted birds perch. In the upper register mares and cows suckle their young while two nomads lying nearby with quivers sew a fleecy sheepskin tunic.

A third person intently milks a ewe with tightly curled fleece while a fourth figure, sculpted like the others in the round, seals an amphora. This scene of domestic life is flanked on either side by a pig, a goat, a kid and a bird. The lower zone of the figures deals with the themes central to the precarious existence of the steppes where only the fittest survive. Three pair of griffins with great feathered wings tear the backs and necks of three powerless horses. This poignant scene leads into sculpted lions and tigers sinking their fangs into a deer on one side and a boar on the other. At the tapering ends of this band, sleek hounds chase hares and pairs of grasshoppers confront each other.

Another astonishing and beautiful example of Greek 4th century B.C. workmanship is seen in a gold comb found in the Ukrainian Soolkha kurgan called "the Witch's mound." The perfectly modeled teeth of the comb, which is five inches wide and five inches high, are topped by a frieze of five miniscule recumbent lions. Above them is a battle group which creates the illusion of three-dimensional sculpture in miniature. The Greek craftsman has treated barbarian subject matter with perfect mastery of the material. Three warriors in the Scythian dress of baggy pants and Greek armor, one of them mounted on a rearing horse, engage in furious combat above a slain horse lying on its back as blood streams from the fatal wound.

A great deal of what we know of the Scythians' appearance comes from such Hellenic pieces as the comb and a gold vessel depicting the nomads in such activities as conversing, stringing a bow, treating a mouth ailment and applying first aid to a comrade's wounded leg. On this and a few other pieces the Scythians are shown with full, luxuriant beards and moustaches and almost never without their characteristic shield, bow and the gorytus, a combination quiver and bow case. Herodotus tells that these were generally made from the tanned skin of an enemy.

Motifs and themes of the rich oral tradition of the Scythians, Samaritans, Alans and Huns were prominent in nomadic art spread westward with the migrations and surfacing not only in the ballads and epics of early Medieval art but also in Carolingain illuminations and gold jewelry and Celtic, Viking and Gothic art and goldwork.

During recent years Russian archaeologists have devoted a great deal of their explorations to the Greek sites along the Black Sea littoral and have been fortunate in discovering a great deal of goldwork of pure Greek inspiration and workmanship. In late 1975 construction workers breaking ground for a building near the coastal resort of Anapa in the northern Caucasus, unearthed a crypt. It dated from the 2nd century A.D. but contained three sarcophagi which had been pillaged in ancient times. Its walls were covered with amazingly well-preserved frescos representing heroes of Greek mythology with great expression and very bright color. Only two meters away was a second crypt with two undisturbed sarcophagi filled with gold crowns and jewelry, silver vases and a great iron sword with a gold-encrusted hilt with a picture of a bird of prey attacking a hare.

(Left) Terracotta figurine of a Carthaginian merchant, ca. 5th century B.C.

(Below) The Carthaginians were highly skilled in the manufacture of glass articles and traded them all over the Mediterranean.

(Left) Etruscan goldwork—a wreath with ivy leaves and berries, a satyr's head at either end.—ca. 400–350 B.C.

(Right) Etruscan goldwork—an earring decorated with bosses, globule clusters, rosettes and filigree, backed with a stamped gold shee— ca. 400–300 B.C.

(Below) Phoenician terracotta figurines of the goddess Tanit, which were manufactured in Carthage, recovered by Bob Marx at Tyre.

(Above, left) Etruscan gold ear-stud with vitreous glass insets decorated with a rosette and concentric bands —530–480 B.C. *(Above, right)* Examples of Scythian gold objects from Kul Oba in the Crimea—4th century B.C.

(Above, left) Scythian goldwork "Kings with dragons"—Tillia tepe Afghanistan, ca. 1st century B.C. *(Above, right)* Greek gold brooch with inscription ca. 450 A.D.

(Above) The Carthaginians dominated the diving for the prized Murex shell from which a valuable purple dye was extracted and sold throughout the ancient world.

(Left) Portrait of Napoleon I on his Imperial Throne, 1806.

CHAPTER 8

GREECE AND PERSIA

The actual amount of gold in the world has been increasing throughout history because gold is virtually indestructible. All, or almost all, of the gold produced since the dawn of time is still with us. Mankind has always had a predilection for confining the shining stuff in the dark—in the tombs of antiquity and subterranean bank vaults of recent times. Consequently gold remains a rare and precious substance which we unfortunately see less of today than did the ancients who gave it pride of place among their treasures.

Through the ages new deposits were continually discovered. Easily gathered alluvial gold was first exploited and then that found underground. Advances in mining and refining have increased yields tremendously by making it possible to extract gold from deep reefs and low yield ores.

Gold, child of the sun and stuff of gods and kings, was the traditional material in which ancient peoples expressed their highest artistic achievement. Gold embodied the peak of accomplishment in the early civilizations of Egypt, Mesopotamia, the Indus Valley, Asia Minor, the Levant and the Aegean. Even in Bronze Age Europe, gold, because it was scarce and associated with religion, attracted the efforts of the most highly skilled craftsmen. Gold in antiquity was drawn as if by magnets to the centers of power and wealth. The mighty states that did not have it beneath their feet expanded frontiers to include gold-filled lands or attracted it from outside sources through trade and conquest, tribute and trickery.

This was true until the millennium before the birth of Christ as gold became increasingly important as a factor in economics. Then things changed and a young nation poor in gold, which expressed its genius in clay, bronze and marble, burst forth with unparalleled flowering of cultural achievements. The city states of gold-poor Greece met the challenge of the mighty Persian Empire which was literally awash with gold and emerged triumphant. For centuries the two contended until the towering figure of Alexander of Macedon broke the Persian giant in the 4th century B.C. With the death of Achaemenid Persia a colossal golden treasure, the world's largest store of gold was released. As the gold flowed west to the new nucleus of power it forever changed the course of history.

The Greeks of the Classical Period anticipated the full range of human potential. Even today Western Civilization mirrors Greek influence in almost every sphere of life—in science, medicine, philosophy, literature, mathematics, education and art. Hellenic Greece lit the world with a many-faceted brilliance still shining today far more richly than the relatively few surviving pieces of gold crafted by the gifted Greek goldsmiths.

Classical Greece matured without the nourishment of gold. Emerging from post-Aegean cultural darkness and poverty, Greece throbbed with the creative intelligence and initiative which peaked in Periclean Athens. But Greece was poor and divided. The city-states, 50 on the island of Crete alone, joined in four unions under the leadership of Argos, Sparta, Athens and Greece and a class of hereditary nobles became powerful. They were often absentee owners of large estates who congregated in the cities to be near the seat of power and the governing Councils on which they served, while goatskin-clad peasants remained poor tilling the soil. These wealthy nobles had the means and time to become proficient well-armed soldiers.

As the Greeks, descendants of landlubbers, became accustomed to the sea they took to coastal marauding. Piracy became common as the sea-roving nobles invaded harbors, looting, burning and taking everything of value. It was during the Age of Nobles that the Black Sea colonies were established. Greek merchants entered in sea-trade using newly built Greek ships based on Phoenician models. These ships sailed to the northern Aegean and to the Black Sea which they called the Pontus. Reaching to the west, Greeks colonized southern Italy, Sicily, Marseilles on the south coast of France and most of the Mediterranean coast of Spain. Gradually the Greeks expanded until their settlements stretched from the Pontus along the Mediterranean's north shore to the Atlantic, while on the southern borders of the Mediterranean Carthage and the Phoenicians vied with each other.

Greek colonial expansion and Greek industry stimulated each other. Greek wares, the elegant and refined products of Greek craftsmen, found ready markets from Etruria to Scythia. And the prospering Greeks at home adopted the idea of coinage from the Lydians. The islands and mainland of Greece quickly took to the system which was an obvious boon to commercial transactions. Greek wealth, previously reckoned in terms of land and livestock, could now be accumulated in coins. With coinage came loans with interest, borrowed from the Near East. The landed nobles were challenged by landless merchants and industrialists whose wealth burgeoned with the introduction of coinage. At the beginning of the 6th century, "Money makes the man" declared the noble Solon, the Greek statesman who had made his fortune in sea-commerce.

But Greek money was based on the silver standard and coined from domestic metal of the rich Laurium mines on the tip of Attica near Athens. Beautiful electrum coins were minted on the Asia Minor coast at the Greek cities of Lampascus and Cyzicus for use on the borders of the Persian Empire where gold was the only acceptable currency. The Mycenaeans had first worked the silver mines in the Second Millennium B.C. In Classical times they were worked by slaves forced to crawl on their bellies in the yard high tunnels, some of which were as deep as 350 feet. Archaeologists have found rings where the slaves were chained to the mine walls. Rarely, in times of crisis when silver could not buy support or supplies, was gold coined in Greece. At the end of the 5th century B.C.

Athens' supply of silver was exhausted by a quarter century of war with Sparta and the city fathers melted down golden offerings to Athena from the Parthenon and made gold coins. Many years later the "debt" to the goddess was repaid with interest in the form of a wooden rack which held the dies with which the gold coins were struck.

Coins played a vital part in the spread of Greek civilization and streamlined commerce in the western Mediterranean, coastal France, Spain, Sicily and in southern Italy. In the Greek colonies of Syracuse and Tarentum gold was mined for coins as early as the 4th century B.C. The Persians, eventually along with most of Southwest Asia, based their currency on gold and the Greeks issued gold coins. Philip II, Alexander's father, issued a huge volume of gold staters bearing his portrait. The 23-karat gold for these coins, which were Europe's first gold currency, came primarily from the rich northern mines of Thrace and Macedonia which had been under Persian control. Philip's staters were so numerous and so widely circulated that they turn up even today all over Europe and Asia. "Phillippus" was an ancient term for any type of fine gold coin and great quantities were drawn to Rome where, in lieu of a domestic issue, they served as the first gold currency following the Roman conquest of Macedonia in the 2nd century B.C.

The fact that Greece had little gold didn't immune the Greeks from its powerful lure. Like every other people they made every effort to secure large amounts of the precious metal. Their widespread trade brought in a certain amount, but until Philip's time Greek silver could not easily be exchanged for gold in trade because the auriferous zones that made Persia rich also supplied abundant silver and dictated a value ratio of 1:13. Egypt, from earliest times, traded gold for silver at a more favorable rate because of her lack of the moon's metal. So it is possible that the Greeks acquired some gold from Egypt even though it was under Persian domination.

The Greek city-states made repeated and vigorous attempts to gain control of gold-producing areas such as the island of Siphnos in the Aegean which also had a lot of silver, and the large auriferous territory lying back of the island of Thasos in the northern Aegean. Such scarcity of gold made it all the more precious. Women craved beauty-enhancing ornaments which they were long denied the permission to wear. Sumptuary laws, restricting the use of gold jewelry, costly raiment, furnishings and accessories had been issued by governments throughout history. They were especially numerous in Ancient Rome but appeared many times in Greece as well, particularly before the 5th century. Plato advocated measures prohibiting the use of gold for personal adornment in his "Laws."

Because poets adored gold, the precious metal belonged more to the world of myth and poetry than the actual world of hard facts and opportunities. It was indispensable in describing the beauty of mortals and the attributes of the gods. The passages of Homer and the hundreds of myths are filled with the gleam of gold. The immortals were credited with having golden parts and some men had the audacity to make similar claims for themselves. Pythagoras, the 6th century B.C. philosopher and mathematician, boasted of his superhuman origin and had a thigh of gold which, Aristotle wrote, he once exhibited to all at the theatre. This was not an original concept. Egyptian pharaohs of the Fifth Dynasty were described as being born with their royal titles inlaid in gold hieroglyphics on their limbs.

What gold the early Hellenes, as the Greeks called themselves, had was considered the property of the state and stored in temples. During the 6th century B.C., known as the Age of Tyrants, the temples of sun-dried brick were replaced with noble temples in limestone or marble. A percentage of the spoils of all battle went to the temples. After the Battle of Plataea in 479 B.C. a part of the Persian gold gathered from the battlefield was made into a golden tripod for the important shrine of Apollo at Delphi. This sanctuary held a portion of the communal gold reserves of almost every Greek state. It was commonly in the form of ingots and could be removed at any time by the depositor. It was also the beneficiary of foreign gifts of gold that awed the Greeks who were unused to such lavish splendors. The oracles delivered by the Greek Sibyl attracted 7,500 pounds of gold sent by Croesus of Lydia. In Philip of Macedon's time Phocian Greeks warring with Central Greece raided the treasury at Delphi, looting the shrine's great treasure. "Then it was," wrote Athanaeus, "that gold blazed up among the Greeks, and silver came romping in."

A certain amount of gold filtered into the hands of the rich whose women at times were able to flaunt their superiority in ostentatious displays of gold jewelry and gold-embroidered garments. The most sybaritic of Greek states in the 7th century B.C. was Corinth whose notorious love of luxury and pleasure influenced the fashions and mores of other Greek cities.

There was evidently enough gold in circulation to make it worthwhile for Periander, the tyrant of Corinth who died in 585 B.C., to go to rather unusual ends to get it. According to the story, Periander needed gold to buy off political enemies, because he announced a great communal festival in which all of the well-to-do matrons of the city were invited. When they were assembled Periander gave a sign to his soldiers who swooped upon the women stripping them of bracelets, rings, necklaces and earrings. They even ripped off any gold embellished robes. These involuntary contributions staved off his adversaries and he continued to rule for another 40 years.

The rivalry among the city-states—resulting in frequent bitter, bloody conflicts that kept them from uniting against common foes—was a boon to the Persians. But the squabbling city-states temporarily overcame their hostilities to repulse the Persians on one important occasion. After the defeat of the oriental invaders at Plataea in 479 B.C. the Greek general, a Spartan, showed his army the gold vessels and plate taken from the royal baggage train of the Persians. "Look at the foolishness of the Medes (Persians). With such provisions for life as you see, they come here to take from us ours which is so humble." His men stripped every dead and wounded foe of his gold and for years afterwards the battlefield was combed for overlooked treasure by gold-hungry Greeks.

Under the leadership of Athens, the Greeks formed the Delian league to drive back the Persian forces and regain the gold-rich areas of Thrace and Macedonia. The defense alliance was named for the sacred island of Delos where the League's treasury was located in the Temple of Apollo. Into its coffers poured contributions from each member, each initially giving according to his ability. Athens, the wealthiest and most powerful member of the confederacy, assessed the annual tribute and soon used this position to force other members into submission. The treasury was moved to Athens by Pericles confirm the city as the center of the Athenian Empire. Pericles was the undisputed leader of Athens from 460 B.C. until his death some 30 years later. He was a brilliant statesman and general who devoted himself to the creation of a splendid Athenian Empire. He envisioned a new and exalted role for the state. To this end he engaged in a 15-year war with Sparta, Athens' chief rival, and used the Athenian navy to blockade the merchant fleet of Sparta's ally, Corinth, ruining its economy.

His fame rests on the fact that during his tenure Athens enjoyed her Golden Age—the era unsurpassed in intellectual, architectural and artistic achievements or accomplishments. Pericles used the proceeds of Athenian extortion to rebuild his city which had been burnt by the Persians in 480 B.C. while the Greek fleet routed the Persians at the famous Battle of Salamis. He restored ancient shrines and public buildings and raised new ones on a magnificent scale. The resources of the empire were devoted to making the capital "in a short time for all time, with a bloom of perpetual newness," as Plutarch later wrote. One of his greatest projects was the noble and beautiful temple dedicated to Athena, the protectress of the Greeks. It was built on the rocky hill of the Acropolis which had first been the site of the citadel of ancient Attic kings and then of simple shrines of sun-dried brick. It took the cream of the nation's architects, engineers, craftsmen, goldsmiths and artists 15 years to create the most gleaming temple which dominated the entire city.

Although it is the marbles that have survived the ages, the influence of work in precious metals on marbles is illustrated by many of the Parthenon friezes. There may have been a variety of gold embellishments as well. A unique gold fragment from the Parthenon, less than three inches high showing Nike in a four-horse chariot, is all that remains of a gold plaque which may have been one of many decorating the temple. The most splendid ornament of the Parthenon was a giant chryselephantine statue of Athena made by Pheidias the famous sculptor and gold-engraver who was Pericles' close friend. This stunning statue was made entirely of gold and ivory—some 2,000 pounds of gold. The Goddess of Wisdom and Chastity, Patroness of the Athenians, towered 38 feet above the marble floor and was placed so that the sun's rays caressed her ivory face and intensified the radiance of her golden crown and flowing robes.

Sometime in the 5th century A.D. the masterpiece was carried off to Constantinople and eventually disappeared so we can only surmise how it looked, although a few small copies have survived. Those who saw the original declared it was the most noble work of art ever created.

The gold was cleverly made of heavy plates which could be removed for any emergency. This proved useful almost at once, for Pheidias and Pericles were accused of having kept a part of the gold earmarked for the statue for themselves. Pheidias was able to have the precious metal removed and weighed proving the accusation false – although one version of the story says some gold was lacking and the two men were fined 40 talents.

In any case Phedias' fame reached the citizens of Olympia who commissioned a statue of Zeus for their city. It was intended as another attraction for the shrine that was the center of the most famous religious and athletic festival in Greece was held every fourth year and attracting thousands of tourists and their money. The Olympian Zeus was even more remarkable than the Athenian statue. For nearly a thousand years the 60-foot high Zeus sat on a great throne of gold, ivory and ebony, paved with jewels. His gold robes were chased with an elaborate pattern of blossoms and leaves. Gold sandals covered his ivory feet and a diadem of golden olive leaves crowned his massive head. In one hand he held a gold scepter inlaid with gems and in the other a chryselephantine statue of Victory. The statue was one of the Seven Wonders of the ancient world—so awesome and magnificent that the great god himself was said to have shown approval by hurling a lightning bolt at its feet.

Pheidias' Zeus also vanished, but archaeologists excavating at Olympia found his workshop there. Following clues in the writings of Pausanius, the same Roman traveler who led Schliemann to the gold of Mycenae, they found the site and some of Pheidias' tools. His two statues were the most splendid but there were others of precious metal or a combination of precious materials. The vestiges of a life-size bull of gilt-silver found at Delphi appears to have been made in one of the Ionian goldworking cities and sent as an offering to the shrine. Large statues of marble inspired by the Minoans were sometimes models for others of clay or stone which were brightly colored and sometimes embellished with gold or gold dust and ivory. A 7th century hymn to Aphrodite evokes an image of one of these statues. "She was clad in a robe outshining the brightness of fire, a fine robe enriched with many colors shimmering like the moon over her tender breasts, a marvel to see. And she had twisted brooches and shinning earrings, flower shaped and round her soft throat were lovely necklaces."

Styles for the scarce Greek jewelry of the 8th century B.C. were strongly influenced by the introduction of Egyptian and Babylonian motifs: the geometric patterns, animal and human figures beloved of Phoenician smiths who opened ateliers in continental Greece and on Rhodes. Persian elements also began to appear in Greek work in gold and silver, particularly in the rhyton and deep bowls such as those of the Achaemenids. The international shrine at Delphi attracted offerings from many distant areas and it was probably there that Greek goldsmiths were introduced to foreign styles. Then, too, there were the Ionian smiths, the Greeks of Asia Minor who were in close contact with the Scythians for whom they executed much goldwork. Some of their production reached Attica and added to the expanding repertoire of goldsmiths there.

The art of the goldsmiths was handed down from father to son and family firms of creators were established which gained widespread fame and grew enormously wealthy. A family on the island of Samos had an association of celators which lasted more than four generations and an association on Chios was called upon to execute works in many distant places.

Until the 5th century Greek women who were excluded from so much of public life were seldom given gold jewelry. Goldsmiths fashioned ornaments and religious objects out of the exceedingly rare substance at the behest of autocrats and noblemen. But in the 5th century wealthy women began to receive a small amount of gold jewelry although the greatest consumers of the still-scarce metal were temples, civic governments, the Graeco-barbarian kings of Asia Minor and sports-loving noblemen.

In the 4th century B.C. when Persian gold stocks were added to the gold from Macedonia, the isle of Siphnos and the Greek colony of Sicily, Greek women were able to indulge themselves in lavish displays of magnificent jewelry. Goldsmiths began turning out larger amounts of jewelry, tableware and religious pieces. Almost none of the tableware or temple gold has survived in its original form. Such articles are always prime candidates for melting down in time of war, hardship or changing fashion. But there are a number of examples of jewelry showing the exquisite taste and skill of the Greek goldsmiths.

They shaped the suddenly abundant metal into breathtaking forms tempered with impeccable refinement of taste. Rings set with cabochon gems, bracelets, diadems, necklaces and elaborate flat mesh chains attest to their talents. They enhanced cast or beaten gold surfaces with raised patterns in repoussé, engraved designs and granulation, enriching the effect with colorful enamels or inlaid stones. The familiar forms from the plant and animal world were translated into shining gold. Bees, acorns, laurel leaves, dolphins and fish were given eternal life. The already immortal deities were modeled into elaborate spiraling bracelets or cast into miniature charms of great delicacy and precision such as the miniscule figure of Eros with upraised feathered wings which came from the atelier of an Athenian smith of the late 4th century.

With the influx of oriental gold more and more individuals wanted to express their status with ornaments and jewelry that had previously been denied them. Consequently, the level of craftsmanship deteriorated as the growing demand was met by many goldsmiths. The master smiths, however, continued to produce masterpieces and train gifted young men to succeed them.

Gold was found in quantity in Greek Sicily. Mines were dug but the seepage of underground water caused many of them to be abandoned. In the 3rd century B.C. Archimedes, the genius mathematician and inventor born in Syracuse, helped solve the problem of flooding. He invented, or perhaps perfected what he had seen in Egypt when he studied at Alexandria, a screw pump for lifting water from the mines. Called the Archimedean screw, it is still used in Egypt and other areas in irrigation and mining.

Vitruvius, the Roman engineer and architect, credited Archimedes with the discovery of the general principle of hydrostatics. According to one story, King Hieron of Syracuse showed him a crown, allegedly of pure gold, and asked if it might not contain an admixture of silver. Archimedes was thinking about this as he stepped into the bath and observed the water overflowing. It suddenly struck him that by putting the crown and equal weights of gold and silver separately into a full vessel of water and noting the variation in overflow, the additional mass introduced by the presence of an alloy could be measured. He was so excited, says the story, that without dressing he ran home to his laboratory shouting "Eureka, eureka, I have found it, I have found it."

In 431 B.C. Sparta, ruled by aristocratic conservatives, declared war on the democratic Athenian Empire headed by Pericles, which embraced the coast of Asia Minor and the Aegean islands. Spartan mistrust of Athens sparked the bitter, demoralizing war which set Greek against Greek for almost 30 years and ended in tragedy as Athens lost all her foreign possessions, her once invincible fleet, and saw the sun set on her Golden Age. As the golden robes of

Athena were removed for war, the Persians took advantage of the civil war to take over the gold mines of northern Greece.

"There is a saying," noted the playwright Euripides, "that gifts gain over even the gods. Over man, gold has greater power than 10,000 arguments." Over the centuries many Greek city gates were swung open by traitors who couldn't resist the gleam of Achaemenid gold. The Greeks too made occasional use of gold to persuade. A northern Greek, Philip II of Macedon, boasted that with a donkey load of gold he could take any citadel. Philip was a military genius and able diplomat who took control of all the city-states of Greece by exploiting their disabling rivalries. His rule brought new stability and prosperity to Greece. From old mines and newly opened ones came gold he coined into the famous Philips. He established new cities, stimulated trade and dedicated himself to crushing the Persian Empire of the Achaemenid kings. In 336 B.C. he readied a united force to invade Persia. But, Persian gold bought a treacherous noble, one of Philip's own men, who slew the Macedonian as he camped with his armies on the shores of the Hellespont. His son Alexander, king at age 20, was left to carry out the mission.

The Delphi oracle told Alexander, "My son thou art invincible." With these words ringing in his ears Alexander crossed the Hellespont into Asia in 334 B.C. at the head of 30,000 troops, 5,000 horses and an official gold prospector. He soon defeated the Persian host amassed to meet him. From the battlefield he gathered 300 magnificent suits of armor richly embossed with gold. The youthful king sent them hack to Athens with a confident message expressing his contempt for the despotic Persians and the Spartans, the only Greeks who had refused to join Philip's league of Corinth. "Alexander, son of Philip, and the Greeks, except the Spartans, have won this spoil from the barbarians of Asia."

In the next 11 years of constant warfare the mighty conqueror won an empire of more than a million and a half square miles and all but monopolized the world's gold supplies. His epochal campaign, during which he never lost a battle, took him over 22,000 miles across mountains, plains and deserts, until he reigned from

Egypt to India. He founded new cities everywhere, mapped unknown lands and opened new worlds to the west. Alexander ended Persian control of the trade routes to the Orient when his troops reached India. The masses of Achaemenid gold along with countless oriental luxuries and novel ideas were sent hack to Greece where they altered first the Greek world and then the Roman. "Then rose the sun of wealth with her far flung might" wrote the poet Pindar. The flood of Persian gold was so great that the value ratio of gold to silver which had been 1:13 under Philip fell to 1:10.

Who were these golden Persians who had long oppressed the Ionian Greeks and meddled in Greek affairs and where had their immense wealth come from? The Achaemenid dynasty emerged from two peoples, the Medes and the Persians who founded the last great Oriental Empire in western Asia. Indo-European tribes led by the Medes of northern Iran are mentioned in Assyrian annals of the 9th century B.C. The Assyrians were expansionists who vied for domination of Western Asia. Assyria briefly held sway over Egypt, and Sargon II boasted that seven Greek kings of Cyprus paid him tribute in gold. In the 7th century under Ashurbanipal Assyria was the world's largest empire, stretching from the Nile nearly to the Caspian Sea and from Cilicia to the Persian Gulf.

The Bible condemns the Assyrians as brutal warriors, armed with iron weapons. Their own annals and artistic records confirm the Hebrew authors. Assyrian kings recounted with pride the wholesale destruction of cities and populations, the grisly tortures and frenzied bloodletting. Bas relief scenes from the great royal palace at Nineveh depict the king's gardens where the trees are hung with the heads of enemy kings. The Assyrian chronicles also recite the rivers of gold, silver, furniture inlaid with gold, gems and slaves that flowed into the palaces, and temple treasuries dedicated to the great sun god Ashur.

Eventually the tribal Mede people balked at payment of tribute and shook off the cruel Assyrian yoke. The Medes managed to subject the Persians to the south around 670 B.C. They established a magnificent capital at Ecbatana, modern Hamadan, whose splendors and rich treasury were described by classical authors. Late in the 7th century B.C. Cyaxares, the Medean king who had been a Scythian subject for almost 30 years, organized the tribal units into a formidable fighting force. The united Medeans joined with the Babylonians who had recently revolted from Assyrian rule to crush Assyrian might. With the aid of Scythians the Medes and Babylonians sacked Nineveh in 612 B.C.

Ancient sources describe how the Assyrian capital held out for almost three years, protected by its monumental double walls and surrounded by a deep moat eight miles in circumference. While the enemy besieged his city, Sardanapalus, the last of 30 Assyrian monarchs, by turns implored his subjects to resist and besotted himself with the debauchery that had characterized his rule. When he saw there was no hope, he gave each of his children 3,000 talents of gold and sent them away to fend for themselves.

Inside the palace he had slaves build a pyre 400 feet high, topped by a fragrant cedar wood chamber 100 feet long. Inside were placed his treasures, the fruit of generations of conquest — gold vessels, gold jewelry, gems, ornaments of many styles and furniture from those countries who had paid homage to Assyria. He had, according to the historians, 150 gold tables and an equal number of gold couches piled in the chamber. On top of the tables were ten million talents of gold and ten times as many of silver. The luxuries of all the world were crammed into the chamber — purple robes embroidered with gold, precious incense and perfumed oils, resins and ointments, rich garments and carved ivory. Then the king followed by a retinue of concubines mounted the wooden stairway to the treasure-filled chamber and gave the order to light the fire. Eyewitnesses reported that the blaze darkened the sky with smoke for 15 days during which the people of Nineveh, unaware of their king's fate, believed him to be offering sacrifices for the sun god's intervention.

Nineveh was obliterated and Assyrian might crumbled. The Medes and Babylonians divided her empire and her gold. Babylon took the southern lands and under Nebuchadnezzar and his Chaldean successors enjoyed a brief period of power and brilliance. But religious conservatism and sybaritic indulgence fed the internal rot which brought the city, one of the seven wonders, to its doom. As its king feasted from the golden plates looted from the Temple at Jerusalem, the Medes and the Persians, under the leadership of Cyrus the Great, entered Babylon without a fight in 539 B.C.

Even before the gates of Babylon were opened to Cyrus, two kingdoms rich in native gold, Lydia and Phrygia, were absorbed by the fledgling Persian Empire. Phrygia was a prosperous country in west central Asia Minor noted for the fertility of its lands and cattle as well as its fine gold. The capital, Gordium, is the seat of King Midas of legendary fame. Until recently he, like so many others, was thought to be a purely mythical figure, but archaeological excavations at Gordium revealed royal tombs and the existence of a dynasty of kings alternately named Gordius and Midas. Legend tells us that the first Phrygian ruler was poor King Gordius whose assets were limited to a pair of oxen. Midas, his son, was a simple but generous soul who had once offered hospitality to a stranger who proved to be the foster father of the god Bacchus. To thank him Bacchus granted Midas one wish. Without reflecting on the possible consequences Midas requested the Golden Touch and it was granted.

Midas ordered an extravagant banquet to celebrate his good fortune and was amazed to find his food turn to gold lumps as it touched his lips. Under his loving caress his cherished daughter turned into a golden statue and the king soon realized that the blessing was a curse. He implored Bacchus to remove it. "Go and wash yourself," commanded the god, "in the waters of the Pactolus." Midas followed this order and the Golden Touch passed from the king to the river. From then on the sands of the river were laden with gold says the Greek legend. This tale reflects the awe in which the gold-poor Greeks held the Pactolus River, a major source of Asia Minor's gold.

The actual location of the fabled river has been lost over the centuries but its riches were real enough in Greek and Roman times. It is believed that the alluvial gold washed down from Mount Tmolus in the Anatolian highlands no longer gilded the river by the Imperial Roman era because all of the gold-bearing quartz matrix had been eroded by the force of water rushing down the steep mountain stream. Not only was Phrygia enriched by the Pactolus but also by gold from mines near Troy and at Lampsacus on the Dardenelles, and most likely from other distant sources through trade. Although their kingdom lay 200 miles inland they had trade tied with the Ionian Greeks and King Midas, married to a Greek princess, was the first foreigner to make an offering to the Oracle at Delphi, which soon began to attract an international following.

Lydia, another gold-rich state to the south, shared the abundant gold of the Pactolus until early in the 7th century B.C. when nomadic Cimmerians invaded the area and overthrew Midas. The king committed suicide as the barbarians, flushed out of southern Russia by the Scythians, assaulted his palace ramparts. Midas' chariot remained at Gordium, according to ancient authors, in the main temple, its shaft tied with a knot of great intricacy. Whoever could untie the knot, said an oracle, would become the Lord of Asia. In the 4th century B.C. a Macedonian general, scarcely more than a boy, took one look at the complex knot, and, as his army watched, slashed through it with a single stroke of his gleaming sword. Alexander the Great fulfilled the oracle and went on to become the Lord of the World.

Cyrus was the son of a Medean princess who married Cambyses I, one of the kings of the people who had settled in the foothills of the Bakhtiari Mountains of Iran. He defeated his own father-in-law to become heir to the Medean Empire and the first of nine Achaemenid dynasts. When he died in battle against the nomads of eastern Iran, he had launched an obscure tribe on a titanic career. In 20 years he had forged an empire stretching from the Mediterranean to the Indus River. In a meteoric rise the Persians threw their snare over one nation after another until their empire stretched from the Black Sea to the Eastern Mediterranean and across into the northern borders of India. Meticulously they subjugated all the world's known gold-producing regions, except for Carthaginian Spain and a few restricted zones of Greek production. Never before had the world seen such golden majesty as in those years when rivers of golden tribute poured into the treasuries of "the Great King, King of Kings, King of the Countries Possessing many Kinds of People, King of this Great Earth Far and Wide." When Darius the Great sat enthroned in his magnificent new palace at Persepolis, ambassadors of some 30 vassal states knelt in homage as they proffered golden tribute.

Darius who ruled from 522 to 486 B.C. issued the first coins with a portrait—his own. These darics were enhanced by the likeness of the King of Kings whose name was respected throughout the ancient world. His coins have been discovered in tombs and ruins from the amber shores of the Baltic to Africa and throughout Central Asia, an indication of the important role they played in trade. The Persians deemed the coinage of gold a royal privilege but allowed subject cities and states to mint silver. Vast numbers of coins were issued and the Persian rulers maintained the integrity of their coinage unlike some of the Greek city-states which began early to debase their gold coinage with silver and copper alloys in a vain attempt to stretch their resources. Shrewd merchants spotted watered coins and raised their prices accordingly, creating inflation that has plagued mankind to this day.

Only four years after the death of Cyrus his son Cambyses II conquered Egypt and Persia and became master of the entire civilized East from the Nile delta at the eastern end of the Mediterranean to the Aegean and eastward almost to India. A mere quarter-century had elapsed since Cyrus overthrew the Medes. Persian conquest was motivated by the lust for gold, the metal sacred to the Persian Zoroastrian religion. Persian rule was marked by military brilliance, effective administration and an economy based on gold currency. Gold brought power and glory to the former shepherds and herdsmen. It bought friends and sowed dissension among enemies.

The Persians loved gold less for its beauty than its potency as a force in politics and economics. The nations which the Achaemenid despots brought to heel were ruled liberally, allowed to keep their customs, religions and languages as long as they produced the gold tribute to fuel the empire. Since the empire was so vast and the Persians were greatly outnumbered, they wisely divided the conquered territories into satrapies or provinces whose chief purpose was to supply gold.

Gold, both raw and worked poured into the royal treasuries at Scbatana, Susa and Persepolis—the most magnificent of the capitals. Even Punic gold reached the Persians through the conquered cities of the eastern Mediterranean which served as depots for Carthaginian gold. The ancient Royal Road from Susa to Sardis and thence to Ephesus, which the Persians took over when they conquered Lydia, had been improved since Croesus' reign. It was the chief artery for caravans laden with the golden fruits of distant mines, booty and tribute to add to the piles accumulating in the treasuries.

The Persians used a lot of gold, a tremendous amount by any standard, for ornament, and ceremonial objects. Their craftsmen were highly competent, and foreign goldsmiths were imported to serve the Achaemenids. The metal's beauty was always secondary to its role in government because Persian omnipotence depended on the gold bars stored in the treasuries. Most gold reaching the capitals, whether in its native state or already handcrafted, was melted down into anonymous ingots and coined as needed to finance the complex and costly bureaucratic machinery of the colossal empire. A tremendous amount went to pay for the constant military operations a huge empire was forced to carry out both in conquest and to maintain power over restless subject states.

The Persians were so enamored of their gold that they even took it into battle with them. Darius II, the weak and cowardly king who was the last of the Achaemenids, rode into battle against Alexander in a golden chariot. He had with him a great baggage train of gold to give him confidence, as well as his queen, who was also his sister, his children and the queen mother Sisygambus. He led his armies with its sea of faces of every hue against the small Greek force, but when expected victory turned to Persian rout, the King of Kings fled the battlefield, abandoning his soldiers, family and much of his gold.

Alexander looked about Darius' palatial tent after the encounter. "So this is what it means to be a king," he reportedly remarked as he surveyed the gold-sheathed and gem-decorated furniture, the golden hangings of oriental splendor and the throne and bath, both of gold. Alexander, who had been taught by Aristotle to love science, philosophy, Greek art and literature, as well as warfare, was not immune to gold fever. As one Persian province after another yielded its gold to him he adopted the manners of an oriental despot and surrounded himself with golden opulence. Damascus, Sidon and Tyre gave up their gold to him. At Sidon there was so much gold that he is said to have had left much of it buried and abandoned. He took golden Egypt and was hailed as king of Babylon as he entered the city, the winter capital of the Persian Kings, in a gold-plated chariot.

He became the King of Kings, King of all the Earth far and near as he swept the world at the head of his great force of fighters attended by thousands of grooms, artisans, merchants, philosophers, doctors, gold prospectors, botanists, engineers, speculators, women and children. He drew from the coffers that fell to him to pay bonuses to his mercenaries and reward his favorites with extravagant gifts. At Babylon he even gave the gold to the priests of Marduk to restore the Tower that Xerxes had destroyed. He could he generous to a fault and delighted in showering his famous "2,000 Companions" with golden victory crowns, coins and confiscated jewelry.

Indulging his flair for the theatrical and to merge Persia and Macedonia into one race, he married Darius' daughter, the Princess Barsine, and at the same time wed 80 of his favorites to aristocratic Persian maidens and regularized the unions of some 10,000 common soldiers with their Persian mistresses. Chares, the court chamberlain, left us a vivid description of the nuptial festivities. A huge pavilion, 800 yards in circumference, was erected supported by 30-foot high gilded columns set with jewels. Carpets of gold and golden draperies, purple and scarlet, decorated the pavilion. Alexander furnished bridal chambers for each of the couples with silver-plated couches; his own was covered with gold. During the celebration, which lasted for five days, he received embassies from vassal states that had once paid tribute to Darius, and they presented him with gold crowns worth an amazing 15,000 talents.

The Conqueror had provided dowries in gold to the newlyweds and this money was soon spent in the marketplace and taverns of Susa. When the Greek soldiers incurred debts and the merchants and hostlers of the city complained, Alexander offered to settle all accounts. Although it cost him 10,000 talents, it endeared him even more to his men who held him in awe. Alexander took on many of the trappings and attitudes of oriental despots, however, much to the discomfort of his troops who felt that the golden luxury of Persia was fogging their leader's head. He even attempted to impose Persian court ritual, which involved prostration in the presence o the king, but the Greeks and Macedonians refused to abase themselves and made snide comments until Alexander astutely abandoned these efforts. His magnetism and his genuine concern for his soldiers, thousands of whom he knew by name, were such that in spite of a volatile temper and his affinity for Eastern absolutism, his men stayed and fought at his side.

In the royal complex at Susa, the summer capital of the Persian Kings, a bonanza of gold and treasure was found by the conquerors. Some sources estimate 500,000 gold darics and such a wealth of jewelry that it was never reckoned. And yet, of it all, the only thing that Alexander is said to have kept for himself was a superb gold casket spangled with gems. He chose it to store his most precious possessions—the copy of the "Iliad" which Aristotle, his childhood tutor had edited for him. After crossing the Hellespont, he had left his armies to make a pilgrimage to Troy, scene of the exploits of his legendary hero Achilles.

The greatest treasure of all came from Persepolis. On other occasions he had been able to hold his armies in check when cities were taken, but at Persepolis he gave them one day free during which they ran amok, looting, raping and slaughtering. Many captives were killed for sport by soldiers already so laden with gold they didn't bother with ransom. The Persian treasurer, a bureaucrat par excellence, who managed to keep the palace vaults intact, turned them over to the invaders and was made governor.

Alexander himself figured the haul from the ceremonial capital at three times that of Susa. There was some two million pounds of gold and silver bullion and specie in the vaults. Before avenging Xerxes' burning of Athens a century and a half earlier, by setting fire to the huge Persian palace the Greeks stripped the city and its inhabitants of every imaginable kind of prize. As Alexander marched away from Persepolis he took with him the treasure of generations of opulence. Plutarch wrote that the vast treasure burdened 20,000 mules and 5,000 camels. Beasts and wagons were piled high with glittering cargoes of gold in the form of ingots, coins, jewels and tableware. There were large numbers of solid gold swords and countless other fantasies such as lustrous robes and tapestries encrusted with gold plaques, embroidery and gems. There were furniture and accessories of precious metals, rare woods, ivory and other exotic materials—luxuries from every corner of the world.

Very little Achaemenid gold has survived but archaeologists sifting through the ashes and rubble at Persepolis found gold and silver tablets bearing the dedicatory building inscriptions of Darius as well as a number of large drinking vessels called rhyta, which combine a conical or horn shaped bowl with a three-dimensional animal figure. Both the bowl and beast, often a winged lion-monster, were hammered up from sheet gold—a very challenging technique—and joined with an invisible seam. The surfaces were richly textured and sometimes enhanced with bands of twisted gold wire, so fine that more than 100 feet could be added to a rhyton without overloading it. Some of these drinking vessels were simply conical solid gold cups in the form of an animal's head such as a gazelle or wild goat with swept back horns. In addition to rhyta of gold and silver, some have turned up made of electrum, the poison-detecting metal prized by ancient rulers. One amazing find was an enchanting bowl of highly stylized flower pedals and stem shapes. Around its eight-inch rim in three ancient languages was the name of its august owner—Xerxes.

Achaemenid art and architecture were cosmopolitan in nature reflecting the imprint of the many ancient people whose lands were subject to the King of Kings. Sculptors, goldsmiths and architects consolidated elements from Egypt, Babylonia, Assyria and the northern areas where the animal style predominated. All of their efforts at artistic synthesis celebrated the divine majesty of the Persian kings.

The greatest monument was Persepolis where stonemasons, engineers, goldsmiths and craftsmen of all kinds and from all parts of the empire labored for generations to create a most splendid complex. Cyrus the Great constructed a 32-acre earthwork platform on a majestic mile-high plateau, 40 miles south of his first capital at Pasargadae. On this terrace, twice the size of the Acropolis, Darius and succeeding kings built a series of vast pillared audience halls, treasuries, harem palaces, banquet and military facilities.

This citadel was the setting for the magnificent New Year's festival, held at the Spring Equinox when the privileged of the kingdom and ambassadors from subject states rode their horses up two flights of broad stone steps to the upper terrace where they paid homage and tribute to the enthroned emperor. Today thousands of sensitively modeled bas relief figures march in silent procession amidst the broken pillars and ruined halls. The poured gold and colorful enamels that once glazed the friezes have long worn off but the sumptuous spectacle is still impressive. Delegations of tribute bearers from 23 vassal countries proceeded along with distinguished Persian and Medean nobles and guardsmen toward the King who sat in full regalia, a golden scepter in his hand. The variety of costumes and faces of the people who lay their gifts at his feet is amazing. The richness of their offerings is staggering.

The greatest cache of Persian treasure was found near the Oxus, known today as Amu Darya, in Turkestan. In 1877 a large number of varied gold objects were offered for sale, one by one, by local peasants. The villagers refused to divulge the location where the treasure had been discovered to the Moslem traders who bought the gold from them. Eventually, through the untiring efforts of an English traveler, Captain F.C. Burton, the hoard was assembled and is now on display in the British Museum. It appears to have been treasure belonging to a temple buried in the 5th or 4th century B.C. to protect it from invaders. Included in this treasure cache were a number of open bracelets with finials in the form of real or mythical animal heads, armlets, necklaces, a sword scabbard, hair ornaments, a variety of cast gold heads and many other objects not as sophisticated as the goldwork of Metropolitan Persia. A gold chariot drawn by four horses formed of sheet gold was modeled with the large wheels used on the Royal Road, the stones of which still bear ruts. There were many interesting rectangular plaques etched with scenes from Persian life showing warriors, heroes, servants and foreign emissaries paying homage.

Where did all the gold come from? While the Greeks were still regaining their balance after the centuries of privation, scattered civilizations in the future domain of the Achaemenids were enjoying gold. Archaeological finds in a number of zones, particularly south of the Caspian, have shown the versatility and imagination of these early Iranian smiths. Very elaborate goldwork was excavated at Marlik representing several distinct styles with some of it made elsewhere. It includes a splendid gold beaker with winged bulls convincingly natural heads and a small gold bowl decorated with falcons and sheep whose bodies are embossed on the gold with heads that protrude from the background.

From Hasanlu, southwest of Lake Urmia, a site connected with the Mannians, comes gold of about the same period, a thousand years or so before Christ. One piece is a large gold bowl embossed with mythical scenes that was found under a skeleton of a man apparently carrying it off when the citadel of Hasanlu was destroyed. Gold goblets, pectorals, bowls, swords and jewelry have come from other sites as well, all of which predate the rise of the Achaemenids. A lot of it is Scythian gold, such as that of the Ziwiye Treasure. Once the Persian Empire was consolidated, gold came from old sources as well as those exploited on a large scale for the first time.

The most ancient supplier of all was Egypt which was easily taken by Cambyses, the unstable son of Cyrus the Great. In its weakened condition the Nile Kingdom produced a great deal of gold. Four centuries later, just before the Roman conquest, the Ptolemies still reaped some 40,000 pounds of fine gold a year. In addition to the traditional workings, Cambyses was attracted to the great gold deposits of the Ethiopian deserts and the island of Meroe in the Upper Nile. Ethiopia was a mysterious land from which distorted tales filtered back of monstrous men with dog faces and others with no nostrils. The Egyptians believed the Ethiopians, slender and handsome, lived incredibly long lives because they consumed only meat and milk. They entombed their dead in great crystal pillars and had endless amounts of gold at their disposition according to contemporary Egyptian reports.

The Persians under Cambyses found no monsters and no crystal pillars, but they did find lots of gold and exacted a terrible toll. Cambyses personally led a gold-seeking expedition into the trackless waste where his men became hopelessly disoriented, crazed with heat, thirst and hunger. They ate all of the pack animals and then drew lots. Every tenth man was slaughtered to feed the survivors. Most of the Persians perished but Cambyses managed to find his way back to Persia where he drowned his awful memories in golden ease. After a few years, he became insane and died.

Gold came from the Hindu Kush and India too, the remotest province of Persia. Herodotus wrote that the products of India included a kind of superior wool growing on trees (cotton) and gold dust which was sent in huge amounts as tribute to Persepolis. One of the richest gold areas paying tribute to Persia was the extensive district comprising Bactria on the northern slopes of the Hindu Kush as far as the Oxus (Amu Darya) River where the giant ants were said to do the digging, and Sogdiana, the mountainous region between the Oxus and Jaxartes (Syr Darya). The Greek Megasthenes wrote that gold came from a mountain plateau bordering modern Kafiristan and Tibet. The furry burrowing ants may conceivably have been the dogs of ancient Tibetan miners or even the miners themselves who worked when the ground was frozen to forestall cave-ins. The popular Tibetan word for the gold collected in the area is "pippilika' which means "ant gold."

At the eastern end of the Black Sea, gold was found in Armenia's Phasis River which rises in the gold-filled Caucasus Mountains. This was probably the same gold Jason sought. In early Roman times Strabo wrote that the mountain torrents carried down large amounts of gold that was gathered by shaggy red-headed, blue-eyed nomads called the "Live-Eaters." "These barbarians," he wrote, "catch it in troughs perforated with holes and in fleecy skins." This description and location jibe with Jason's quest for the Golden Fleece of Colchis.

The Persian Empire was far more extensive than any that preceded it and the official lists of the subject territories which rendered tribute include 30 provinces. The gold they anted up worked miracles for the Persians who used it to corrupt and divide their enemies. Once Alexander liberated the truly colossal volume of gold it began to filter through his armies into the marketplaces locally and then in Greece where it tipped the balance of history. The uniform coinage Alexander instituted for his great empire brought a new prosperity to Greece, and during the three centuries following his death the Greeks of the Hellenistic Age extended the influence of the most remarkable secular figure in ancient history.

When Alexander died of fever in Babylon in the summer of 323 B.C., he was not yet 33 years old. He had been driven by all-consuming ambition from Egypt to India, sharing every hardship with his faithful men. He embodied the Greek ideal of living life to the full. Like Achilles he welcomed a short life but one brimming with glorious deeds. "It is a lovely thing to live with courage," he once said to his assembled troops, "and leave behind an everlasting renown." After his death his generals, the diadochi, embarked on a bitter power struggle that tore apart the empire in far less time than it had taken to build. Three dominant powers emerged—the Egypt of the Ptolemies that was to maintain Hellenistic rule until Cleopatra's death in 30 B.C.; the Selucid Empire in Syria; and the Macedonia of Antigonus.

As Alexander lay dying, a workshop was set up where the most outstanding goldsmiths of the world labored on a sarcophagus and funeral equipment worthy of the "World-Seeker." An eyewitness reported that "It was more magnificent when seen than when described, the most splendid funeral display the world had ever seen." Sixty-four mules with gilded crowns and trappings, gold bells and collars studded with gems pulled the huge gold shrine. Within the jeweled temple was a coffin of beaten gold enclosing the body of Alexander preserved with spices. Week after week, month after month the catafalque progressed, crossing a thousand miles, reverenced in every village and settlement. Ptolemy managed to divert it to Egypt where it came to rest at Alexander, the crown jewel of Alexander's 70 cities. There the golden coffin remained, the object of adoration and sacrifices, until the degenerate Ptolemy IX melted the sarcophagus three centuries later to strike coinage to meet an army payroll. When the coffin was opened, Alexander's face, 300 years old, was clear and handsome as in life.

A legend still persists, one of thousands that grew up about Alexander, which concerns buried gold. It is said that when Darius III fled towards Ecbatana he took with him huge amounts of gold and that when he was murdered it could not be found and was presumed to have been buried somewhere around Ecbatana. The king also had all of the soldiers involved with burying the gold silenced by death, leaving no one to reveal the treasure's location.

Since that time Persians, Parthians, Greeks and Romans have all searched in vain for Darius' gold. The Romans were particularly tantalized by the Ecbatana golden treasure. When the Triumvirate of Pompey, Caesar and Crassus divided the world in three parts in the century before Christ's birth, Crassus, already the richest man in Rome, headed for Ecbatana. Ignoring the advice of the contemporary poet Horace, who admonished "Be wise—ignore the hoard of gold once hid," Crassus hoped to find the treasure that had eluded Alexander. According to Plutarch, the Roman treasure hunter had a most unpropitious start. As he set out from Rome, one of his political opponents placed a burning blazer at the city gate, casting incense on the flames and invoking curses of so dreadful a nature as Crassus appeared that no one believed he could avoid disaster.

At the Adriatic port of Brindisi, where he was to sail for Asia, an old fig seller warned him not to embark. But gold fever possessed the Roman and when the voyage began, he overruled the galley captain who urged a southerly route because of bad weather. Gold-hungry Crassus insisted they go by way of Carrhae on a direct route for Ecbatana. At Carrhae, nemesis caught up with Crassus in the form of a large Parthian army. After his capture, the enemies, aware of his gold lust, amused themselves by pouring molten gold down his throat. His gold-filled head was sent as a trophy to the Parthian monarch as he was attending a performance of Euripides' "Bacchae." The king threw Crassus' head to the players, saying that they might keep the gold if the performance was good.

Other Romans followed in Crassus' wake looking for the elusive Ecbatana gold and were also unsuccessful. One of the most notable attempts was allegedly made by the emperor Trajan. Believing the vast treasure to be buried in a river bed, he dispatched a great number of laborers to excavate a new channel to divert the river and allow the treasure to be found. However, this imaginative effort revealed no glint of shining gold. Darius' gold is still being sought today.

Persian King Darius III and Persepolis frieze of Libyan or Syrian tribute bearers.

(Left) Tile fresco depicting Alexander the Great.

(Above) Coinage of King Lysimachus of Thrace—gold stater depicting the head of deified Alexander the Great adorned with the ram's horn of the Egyptian god Ammon. On the reverse is a seated figure of the goddess Athena, ca. 300 B.C.

(Above) Gold Stater of Philip II of Macedon—the father of Alexander the Great—ca. 359-336 B.C.

(Left) Persian gold cup depicting mythical beasts, ca. 2400–2300 B.C.

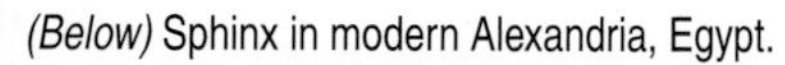

(Below) Sphinx in modern Alexandria, Egypt.

Modern Alexandria at night.

(Above) Ruins of Tyre, Lebanon, conquered by Alexander the Great in 332 B.C.

(Left) The fantastic remains of the Triumphal Arch in Tyre, built in honor of Alexander the Great in 332 B.C.

(Left) These two Greek amphorae date from the time of Alexander the Great's conquest. Bob Marx recovered them off Tyre.

(Below) Turkish Koran Case made of gold and ornamented with diamonds and pearls, 19th century.

(Right) Gold aureus depicting Nero and his mother Agrippina, ca. 54 A.D.

(Below) Roman rings, two with Egyptian scarabs, ca. 4th–3rd century B.C.

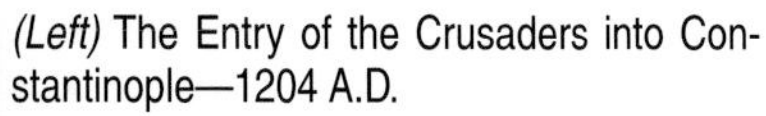

(Left) The Entry of the Crusaders into Constantinople—1204 A.D.

(Below) Aureus minted in 193 by Septimius Severus to celebrate the legion that proclaimed him emperor.

(Far left) 16th Century Turkish gold pitcher decorated with many precious stones.

(Near left) Turkish decorative hanging of gold and precious stones from 18th century.

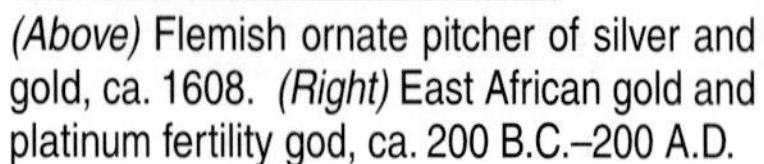

(Above) Flemish ornate pitcher of silver and gold, ca. 1608. *(Right)* East African gold and platinum fertility god, ca. 200 B.C.–200 A.D.

CHAPTER 9

AURUM ROMANUM

The modest beginnings of antiquity's greatest empire showed no sign of the golden splendors to come. In the days when the Etruscans landed on Italy's pleasant shores to found their cosmopolitan cities and produce some of history's most remarkable goldwork, Rome was no more than a gaggle of simple farming villages scattered on some hills just south of the Tiber, Italy's only navigable river. When the Italic tribesmen wanted metal implements or weapons they took cattle and grain to a trading post frequented by Etruscan traders who crossed over from the other side of the river. This market, known as the Forum, was held on low marshy ground surrounded by hills.

It was there that Rome began to grow in the 8th century B.C., forming around a village on the Palatine hill which absorbed its neighbors. Tradition credits Romulus, a descendant of the Trojan hero Aeneas, with establishing the city in 750 B.C. But despite the glory of Rome's alleged foundation, it was in reality a poor city state. For centuries Roman kings, the first of whom were Etruscan, struggled to survive and maintain a viable economy. Etruscan civilization with its Greek overtones gave a great boost to Roman development. Under Etruscan kings the city was greatly improved by ambitious engineering and construction projects. The state prospered and expanded and ca. 500 B.C. the native Latin inhabitants expelled their Etruscan rulers and established the Roman Republic.

Even then, compared with the other Mediterranean powers, Rome appeared a most unlikely prospect for a starring role in history. Fourth-century B.C. Rome was constantly engaged in warfare and continually threatened by invasion the Greeks who held southern Italy (Magna Graecia) and the Carthaginians. The early Republic had an exceedingly small stock of gold, the traditional symbol of worldly success. Barter gave way very slowly to more sophisticated transactions. Increased commerce with the Greek cities of the south led to the issuance of Rome's first coinage, bronze rather than silver which was circulated in the Greek world. The bronze *as*, minted in the second half of the 4th century B.C., weighed nearly a pound troy, and was divided into 12 smaller coins, each called an *uncia* or ounce. Two generations later, in the mid-3rd century B.C., following the capture of the rich Greek cities of southern Italy, the Romans began to coin silver.

Gold remained too scarce to use for currency during the Republic's early days. Rome had an insignificant amount of gold; most of it from Etruscan sources. By 390 B.C. when marauding Gauls threatened Rome, the city fathers couldn't scrape together more than 1,000 Roman pounds of gold to meet the enemy's ransom demand. The Roman pound contained 5,040 grains and was quite a bit lighter than the standard English pound of 7,000 grains. That 1,000 pounds of gold represented the Republic's entire national reserve after more than a century of progress.

The next centuries, however, saw the steady rise of Rome's fortunes with the dawn of the Christian era. The city on the Tiber had become capital of a mighty empire, a dazzling magnet attracting the world's wealth. Gold, particularly the incredible spoils from Egypt and the produce of the Spanish mines, was the single most important factor in the striking change in lifestyle that took place among the well-to-do Romans. There was a shift away from the traditional values that esteemed a devout, sober and modest way of life. The "pietas, gravitas, semplicitas" of a nation of proud farmers were replaced by homage to "sanctissinia divitiarum majestas," the sacred majesty of wealth of which gold was the supreme symbol.

The Romans adored gold but not because of its links with religion or its inherent and eternal beauty. Gold was prized because it bought power and because it was ideally suited as a means of displaying wealth and influence. The Romans were, above all, a practical people. They were not as sensitive to refined beauty as the Greeks, whose art works they admiringly collected. Nor were they as subtle. In classical Greece, gold ornaments of exquisite beauty and elegant simplicity had been worn somewhat sparingly. Grandeur and magnificence in buildings had been reserved for the gods. But in Rome, good taste was generally sacrificed on the altar of ostentation. The excessive oriental tastes of Egypt's Hellenistic kings, which came to Rome along with Cleopatra's gold, influenced Roman styles in jewelry, dress, architecture and manners. Further Eastern conquests reinforced the love of extravagant display.

Women and sometimes men weighed themselves down with gold. Contemporary writers commented on the rage for wearing quantities of ornate bracelets and rings. "Why," Ovid asks a fashionable matron decked out in pendant earrings which reached almost to her shoulders, numerous overworked bracelets, brooches, pendants and rings, "are you so anxious to carry your wealth on your person?" In the "Satyricon" Petronius satirized the vulgar taste of wealthy freedmen who were notorious for their excesses in the early Empire. At a party the hostess Fortunata removes her golden ornaments one by one, commenting on their value.

"It soon got to the point when Fortunata took the bracelets from her great fat arms and showed them to the admiring Scintilla. In the end she even undid her anklets and her gold hair net." Her husband, Trimalchio, comments "A woman's chains, you see. This is the way us poor fools get robbed. She must have six and a half pounds on her. Still I've got a bracelet myself, made up from the one-tenth per cent offered to Mercury—and it weighs not an ounce less than ten pounds." Trimalchio, afraid of being taken for a liar, has servants bring a scale and weighs his bracelet before his guests. Scintilla takes a large gold locket from around her neck and removes two earrings with jingling beads to show everyone. Habinnas, her husband, remarks: "You cleaned me out. Honestly if I had a daughter, I'd cut her little ears off. If there weren't any women everything would be dirt cheap."

"Our parents have trained us up to an admiration of gold and silver," wrote Seneca in the 1st century A.D., "and the love of these metals has grown with us to such a degree that when we would show our gratitude to heaven we make presents to it of those metals. This fact it is that makes poverty seem a curse and a reproach; and the poets help it forward – the chariot of the sun must be all of gold, the best of ages must be golden and thus they try to make the greatest possible blessing of mankind what in reality is the largest cause of misery of mankind."

The epitome of Roman gold lust and decadence was reached by the emperor Elagabalus, who devoted his short reign to the worship of pleasure and made use of gold in ways that would have appalled the hardy, simple peasants who launched Rome. Elagabalus wasn't even Roman. He was a 14-year-old Syrian who had been trained as a priest of the oriental god Baal. His powerful grandmother bought the throne for him in the troubled early 3rd century A.D. and he left the governing to her while he indulged in every kind of excess. The utterly debauched emperor amused himself by tossing gold and gems into a herd of cattle and watching as greedy Romans were trampled to death as they plunged into the milling mass. Another of his pastimes was watching from the Basilica Julia as frenzied crowds competed and killed for handfuls of gold coins he tossed down. He held public lotteries for prizes which typically included a dozen gold pieces, a dozen pearls, a dozen eggs and a dozen dead flies.

In spite of these perverted games, the Roman mob liked their boy emperor. While his ambitious grandmother ruled, he welcomed Romans of all classes to his magnificent palace where the chamber pots were gold and slaves gilded the porticoes with gold and silver dust. He received the public wearing rouge and eye makeup, dressed in cloth of gold, embroidered silks and loaded with gem-studded jewelry. The parasites surrounding Elagabalus were willing to run the risk of being the butt of his warped sense of humor because the possible rewards were great. Sycophants flocked to his bizarre banquets held in gold- and ivory-paneled halls. They reclined on gold-plated couches covered with gorgeous gold embroidered spreads, previously outlawed before Rome's leaders made a mockery of the stern ideals on which the state had been founded.

Elagabalus could be very generous. Prizes were often distributed at the end of an evening's revelry. Lucky banqueters might receive the golden plates they had eaten from, sacks of gold coins, chariots or prize horses. Performers who pleased the emperor were handsomely paid. Those who didn't might receive a dead dog or a prison sentence. Depending on his whim, slaves might bring in platters of peas mixed with gold pieces or serve amber beads in the beans and pearls in the rice. On the other hand, Elagabalus might just as easily have his guests served gold and silver plates heaped with perfectly modeled wax or wood replicas of food. They were forced to watch without comment while the emperor proceeded to consume an enormous meal of such exotic delicacies as camel's heels and flamingo tongues.

He once gave a party of banqueters 1,000 pounds of gold when he lost a bet with them. Such tantalizing prizes kept them coming in spite of themselves. Elagabalus enjoyed drugging guests as they ate. A man might wake to find himself in the arena surrounded by wild beasts, so dazed that it took him some time to discover they were toothless creatures the emperor kept for terrorizing his companions. Sometimes the terror was of a gentler but more deadly nature. A trick ceiling in one of his banquet halls might open to shower down perfumes, gold or flowers. More than one Roman allegedly smothered in a downpour of blossoms or was knocked senseless by a golden shower.

For four years Rome was treated to such excesses. Then the emperor prepared for the violent death that fellow priests of his oriental cult had predicted for him. He declared no one would ever die as luxuriously as he and readied a golden suicide tower where he collected gold daggers and a variety of poisons in containers of hollowed out sapphires. But in the end his plan was thwarted and he died as violently and ignobly as so many of his fellow emperors—slain in an outhouse by members of the powerful Praetorian Guard. His body was dragged through the filthy cobbled streets and flung into the muddy Tiber to join the remains of so many other men whose gold couldn't save them.

Roman history is a tangled saga of struggle and intrigue drenched in blood and spangled with gold. The precious metal was ever the means and the end of power politics. Gold, which was in such short supply in Rome's infancy that the Twelve Tables, the first compilation of Roman law ca. 450 B.C., forbade burying it with the dead became Rome's chief export. The world's gold was drawn to Rome and then pumped from the heart of the Empire through the network of commercial veins reaching as far as Han China, source of the precious silks coveted by elegant Roman ladies.

Gold was essential for buying power at home and luxury goods from abroad. The Romans had some industries but they were not large-scale manufacturers and depended on the production of their colonies and foreign states such as India and China. Over the centuries a complex financial system involving checks, mortgages and joint stock companies developed, but gold continued to be the most trusted unit of value. As the Empire began to slide into bankruptcy and chaos, gold continued to serve as a hedge against inflation. After the 3rd century A.D. when watered currency provoked crippling inflation which stripped men of their savings and decimated property values, gold was often the only acceptable medium of exchange.

At the peak of her glory Rome was mistress of the world. She controlled every gold source known to the West. Gold from Central and Southern Europe, Asia Minor, Egypt, Africa, Britain, and above all from Spain, was the backbone of the Roman economy. Rome's first gold was Etruscan booty, most of which came from the Alpine headwaters of the Po River in northern Italy and some from France. Conquest of the Etruscans must also have provided some gold from Greek and Carthaginian sources. But until Alexander liberated the colossal volume of Persian gold and sent it flowing westward, there was relatively little gold to be had.

As Rome embarked on its vigorous program of territorial expansion in the 4th century B.C. securing new supplies of gold became imperative. Gold washed down from the western Alps couldn't begin to satisfy the growing Roman appetite. The object of almost every conquest was either a land blessed with native gold or one with a stockpile of accumulated treasure. The conquered territories were considered Roman property. Native populations were permitted to continue in residence as long as they paid for the privilege. Vanquished states were stripped of all treasure and then subjected to systematic heavy taxation. Since silver was the useful metal in commerce with the Hellenistic Greek world and since gold was regarded as a national reserve for times of crisis, Rome initially exacted tribute in silver rather than gold.

There were strict sumptuary laws, often ignored, discouraging personal use of gold, as it began to flow into the treasury. Gold came from the spoils of battle, war indemnities, tribute, taxation and continued exploitation of deposits. The first dramatic increase came with the conquest of Magna Graecia, southern Italy, from which came a great deal of worked gold. Then Rome, as victor in the First Punic War with Carthage, gained Sicily and war damages of 3,200 gold talents.

With the subjection of northern Greece, Rome gained manufactured gold in unprecedented amounts, and the mines, which Philip of Macedon, Alexander's father, had organized. By 189 B.C. Roman armies had reduced Macedonia and Syria, two of the three great empires which succeeded Alexander in the East, to vassal states. Plutarch described how the Roman general Aemilius Paulus operated in northern Greece. At the capital city of Epirus "he sent for ten of the principal inhabitants of each city and fixed a day for them to bring whatever silver and gold could be found in their houses and temples. With each of these he sent a centurion and a guard of soldiers under the pretense of searching for and receiving precious metals, and as if for this purpose only. But when the day came they rushed upon all the inhabitants (who had previously been disarmed) and began to seize and plunder them. Thus, in one hour 150,000 persons were made slaves and 70 cities sacked."

The slaves were sold to the magnarii, the wholesale merchants who followed the army, for about a dollar a head. The magnarii together with the generals and the joint-stock companies of "tax-farmers" who wrung taxes out of the subjected peoples for the government, keeping a percentage for themselves, profited greatly from expansionist ventures. Whole cities were often sold on the auction block. The entire aristocracy of Macedonia including Greek intellectuals, doctors and bankers were auctioned off at Delos, the chief Roman slave mart which had once been the sacred island home of the Delian League.

The wholesalers accompanied the armies. They often advanced enormous sums to finance campaigns for generals as eager as they to make their fortunes. As long as Rome kept reaching out to conquer new lands, the immediate rewards of battle were great and many men became rich literally overnight. The lion's share of booty went to the generals, but the soldiers received a share too. Men fought for profit rather than the honor of the patria which had motivated the citizen-soldiers of the early Republic. Once common soldiers tasted the gold of conquest they were unwilling to return to the drudgery of their small farms which were oppressively taxed. Moneyed aristocrats bought farms cheaply from war veterans and over-taxed farmers. Later they acquired estates through the proscription lists which condemned an emperor's enemies and made his lands available to the ruler's friends for a fraction of their value. In this way, the fertile Italian plots that were the inheritance of Italic peasants passed into the hands of a relatively few feudal landowners.

Dispossessed farmers and their families flocked to the towns and especially to the glittering capital. There they swelled the urban masses who shared very little in the opulence enjoyed by the patricians, knights and rich freedmen. They lived in appalling slums and were easy prey for epidemics, violence and manipulative politicians. The mob was placated with bread and circuses —handouts of grain and magnificent public entertainments. Julius Caesar began what became the standard practice of giving out donatives, gifts of money to bolster a ruler's popularity. People came to expect these gifts as well as gold for their votes.

The first of the great public spectacles honoring a victorious general was the triumph held for Aemilius Paulus after the Battle of Pydna in 168 B.C. He brought back such an overwhelming amount of treasure that Republican Rome never again had to tax her citizens. It took three days to parade the wealth of the Macedonian court through the streets. The third day was reserved for the gold as 120 sacrificial white oxen with gilded horns and wearing golden wreaths led the procession. Marching children carried gold cups and libation vessels. There were large vessels of beaten gold and gold tableware that had belonged to Antigonus and Seleucis, Alexander's generals. A parade of men carried great containers of gold coins, four men to each of the 77 vessels. There was a magnificent gem-strewn cup made of ten talents of gold, 400 gold crowns and wreaths from fallen cities and chests and chests of gold ornaments and jewelry. Last of all came Perseus, the defeated Macedonian king, and his children in chains.

None of the treasures described by Plutarch, who detailed the triumphal procession, have survived. But a treasure unearthed by chance in northern Greece gives an indication of what the Romans saw. The hoard, found in the area of Domokos, was probably buried during the war between Perseus and Rome. It includes the popular serpent bracelets with elaborate detailing and gems set in the smaller coils near head and tail. There are also a number of gold medallions with busts of such deities as Aphrodite and Artemis, linked to a network of finely wrought gold chain. Among the most characteristic pieces were gold belts with buckles in the shape of the so-called Hercules knot, which was supposed to have magical powers. The intricate knots and the terminals of the chain or mesh links are elaborately decorated with filigree, granulation, enamel, glass paste and inlaid stones. A gold diadem with a Hercules knot features dangling beads fastened in clusters to the knot which is flanked by Ionic capitals embellished with geometric and floral designs. It is outstanding in rich use of color and texture. There are also earrings, funerary wreaths of sheet gold shaped like fruit and flowers, bracelets and necklaces which have been found in other hoards and are on display in museums. These Hellenistic gold works must have whet the appetite of all who saw them.

Rome found itself caught up in a vicious cycle. Once she had committed herself to a career of conquest, there was no stopping. The First Punic War, begun in 264 B.C., had pitted the seafaring Carthaginians, Rome's chief rival, against peasant-warriors with no naval experience. The Romans were determined to rout the descendants of the Phoenicians who had turned the Western Mediterranean into an exclusively Carthaginian sea and controlled the vital Strait of Messina between Sicily and Italy. They built a fleet and trained a navy. When the fleet was destroyed after several naval victories, they built another. One fleet after another was lost until the treasury was empty. The citizens of Rome responded to an appeal for funds and a final fleet was constructed which finally defeated the Carthaginians.

The amazing victory changed the course of Roman destiny, committing her to the perpetual maintenance of a standing army. To afford such a military complex, Rome had to continually expand to seek new sources of booty, tribute, taxes and mineral wealth. These endeavors required, in turn, an ever bigger and better army and administrative bureaucracy.

The Second Punic War began in Spain in 218 B.C. and ended with the defeat of the Carthaginians in Africa 17 years later. This was the war during which Hannibal led his forces and the famous elephants over the Alps into Italy and almost succeeded in defeating Rome on her own ground. After the hard-won victory, Rome exacted 10,000 talents in war damages from Carthage.

This amount was added to the spoils of numerous other victories. Then, Flaminius brought some 5,000 pounds of gold to Rome from his victory in Greece in 194 B.C. The royal treasure of Hiero, the king of Syracuse, fell to Rome as did 2,070 Roman pounds of gold from the city of Capua, over 270 from Spanish New Carthage and some 5,000 pounds from Tarentum. In 210 B.C. the Senate had levied a tax on privately held gold, a sign that gold ownership was spreading. The historian Livy calculated that at the beginning of the Second Punic War Rome had about 4,000 pounds of gold in its treasury, four times as much as had been there in 390 B.C. when the Gauls demanded gold as the price of peace. By the end of the war Rome had a great deal of gold, for the first time in her history.

The greatest prize of the Second Punic War was Spain, the storehouse of treasure which filled Rome's coffers and accelerated the corruption, venality and extravagance which widened the gulf between the rich and the majority of Romans who lived in comparative squalor. Despite the severe financial drain of constant campaigns waged between the end of the Second Punic War and the beginning of the Third and last Punic War in 150 B.C., the Treasury had almost 18,000 Roman pounds in reserve when Rome initiated the war which ended in 146 B.C. with the complete destruction of Carthage.

Spain enriched many Romans. Influential men sought appointment as provincial governor, even though the position carried no salary. They served not out of civic pride but because a year in the post could set a man up for life. Verres, governor of Sicily, told Cicero that a three-year term was more than sufficient. "In the first year," reported Cicero, "he could secure plunder for himself, in the second for his friends, and in the third for his judges." Spain was the plum of such appointments. Caesar went to Spain as quaestor and returned to Rome with enough gold to pay off a monumental debt to Crassus, the greedy Roman who died with a throat full of molten gold in Persia. Crassus, one of Rome's biggest tycoons, had made his first millions as a slum lord. He lent Caesar gold to advance his political career by buying votes and financing public games.

Caesar borrowed heavily from Crassus again as he moved up the political ladder. He repaid a huge debt by going to Spain as governor. He came back so wealthy that even after he paid off his debts, he poured such a huge amount into the treasury that the Senate voted him a triumph. Later Caesar waged a nine-year campaign in Gaul and added territory twice as large as Italy to the Roman Republic. Plutarch insisted his object in the bloody Gallic wars was to gain gold rather than win the population for Rome. The conquest cost an estimated million Gallic dead and a million enslaved. It brought Caesar untold wealth which he used to advance his ambitions, liberally distributing it and financing vast public works projects that employed thousands of poor Romans.

The conquest of Gaul flooded Rome with gold. The ratio of silver to gold, which had been about 15 to one in early Rome, fluctuated somewhat with her shifting fortunes. The market price dropped some 18% when Gallic gold reached Rome, but because it was not yet a monetary metal, the ratio soon normalized and throughout the Empire generally hovered at to 13 to one.

Gold coinage was instituted under Caesar. Before him, the only gold coins minted at Rome were extremely rare emergency issues such as a stater struck during the Second Punic War. Caesar used his coins to pay troops. A common foot soldier received a gold aureus for two weeks service. Other generals used captured gold to pay their men. One of them was Sulla, one of the first dictators who came to power on a wave of war gold and Roman blood. As a general in Greece he looted the temple treasuries of Delphi, Olympia and Epidaurus and sacked Athens. He returned from Asia Minor with 20,000 talents in indemnities and taxes as well as 15,000 pounds of gold and 115,000 pounds of silver for the treasury.

Known as Sulla Felix, because of his sense of humor and love of pleasure, he was born to a penniless patrician. He allied himself with the arch conservative cause before going abroad. When he returned he found Rome a battleground contested by several factions; 10,000 people had been slaughtered in a pitched battle in the center of the city in one day. Sulla's arch enemy Marius ruled through a year of terror during which neither rich nor poor were secure.

Sulla used his 40,000 loyal veterans and his gold plunder to overcome the democratic forces that made some effort to enfranchise the poor and reduce slavery. In 82 B.C. he became dictator and launched a blood-chilling campaign of revenge. His first act was to issue a proscription list calling for the deaths of 90 senators and 2,600 equestrians or knights. He offered golden rewards to their murderers and soon the Forum was adorned with ghastly severed heads. Sulla the Happy offered up huge golden rewards for each head of an enemy or men whose treasures or estates he wanted. During the Second Triumvirate 40 years later, assassins could get even larger payments of gold for each head. Usually, however the price was low and aspiring politicians found it more expedient and often cheaper to have their rivals murdered than to buy favor.

Candidates openly handed out bribes. Cicero describes the handing out of coins on the Field of Mars. Mass bribes and lavish spectacles secured the votes of whole segments of the population. The Senate did little to curb the practice since few lawmakers were above doing the same. The sure winners in even the most hotly contested elections were the temple bankers and the moneychangers whose shops could be found in a crooked row leading from the Forum. Huge sums at inflated rates of interest were loaned to those who weren't fortunate enough to have either a personal fortune or a rich patron.

Caesar's murder in 44 B.C. plunged the faltering Republic into 13 years of monstrous civil war which ended only when Octavian, known as Augustus, emerged triumphant over Mark Antony, Caesar's trusted general. Caesar's will named Octavian heir to three-quarters of his vast fortune and named him the slain dictator's posthumous adopted son. But before Octavian could get his hands on Caesar's gold, Mark Antony managed to remove some of the gold Caesar had on deposit in the Temple of Ops, goddess of wealth. Initially the two opponents joined in an alliance of sorts to avenge Caesar's death. With Lepidus they formed the Second Triumvirate and immediately proscribed 300 senators and 2,000 equestrians. They established a Sulla-like reign of terror motivated as much by the need for gold as for revenge. Without gold they couldn't pay their armies; without their armies they would have no power and without power they could gain no gold to pay their armies. The mere possession of money became criminal. Children who inherited estates were executed and thousands were slain as exits to the city were blocked.

The provinces didn't escape during this period. Brutus and Cassius, opposing the Triumvirate, led armies through the East demanding ten years tribute in advance to support their troops. The population of Cilicia in southeast Asia Minor was sold into slavery when they couldn't meet a ransom demand, even after all the gold of centuries had been collected from temples, public buildings and private homes and melted down. Cassius demanded over four million shekels in gold and silver in Judea and when it wasn't forthcoming he sold the inhabitants of four towns as slaves. Cassius and Brutus were killed in battle in Macedonia, but the struggle continued as Octavian and Mark Antony vied for supremacy. Rome and the provinces suffered greatly until Antony fell victim to the twofold lure of Egypt's famed treasures and the temptress queen who had already beguiled Caesar.

After the suicides of Antony and Cleopatra, following the decisive defeat in the sea battle off Actium, Augustus was proclaimed emperor and Egypt became his personal estate. The Egyptian treasury of the Ptolemies and Cleopatra's private treasures infused new life in the stagnant Roman economy. When Augustus mounted the throne Rome was in terrible shape after the years of bitter civil strife.

Egyptian gold was added to Spanish, Gallic, Balkan and Alpine supplies to furnish the material for Augustus' reformation of the monetary system, one of his first and most significant acts.

He issued gold aureii in great numbers. Money became so abundant that, according to Suetonius, the interest rate fell from twelve to four per cent and the value of real estate soared. With Augustus' affirmation of the right of individuals to amass wealth, the gold that had been in hiding since Caesar's death began to emerge to stimulate trade and investments. Life improved for everyone. Even a man of moderate means could hope to collect a small nest egg and slaves could aspire to buy their freedom, although this seldom happened. The rich, meanwhile, emulated the self-indulgent splendors of the East, rivaling oriental courts in their homes, dress and manners.

From Augustus' reign onward, Roman emperors turned out great numbers of beautiful coins which remain the highest expression of Roman artistry. The earlier military coinage was the model for imperial currency. The emperors had their portraits on one side of a coin and boasted of achievements or announced plans on the other. Imperial coins, most of them of very high quality gold, were often masterpieces of the ancient art of the celator or engraver. They served as propaganda in an era long before mass diffusion of information. Issued from 600 mints in the Western Empire and later Eastern Empire, Roman coins provide a fascinating record of history. The coins traveled throughout the world and have been dug up in modern times from sites in China, India, Africa, Russia and the icy vastness of Scandinavia.

Being emperor was a dangerous occupation. Few died peacefully and fewer still held the throne for long. Between 235 and 285 A.D., for example, 26 emperors were crowned and only one of them enjoyed a natural death. Six men were named emperor in the year 238 AD. Four were murdered or killed themselves and one was slain in battle. Time was of the essence for these insecure rulers whose vanity prompted them to rush production of commemorative coins. Almost every single emperor, no matter how brief his reign, managed to have at least one gold coin made to ensure his immortal fame.

The Romans were great collectors of ancient coins and artifacts.

Counterfeiters "falsarii," if caught, were hanged or thrown to the lions, but there was a tempting market for replica coins. In an attempt to keep the unscrupulous from deducting a bit of gold from each gold coin that passed through their hands by shaving a bit of metal off the edge, the Romans invented the milled edge. But they couldn't stop the practice of lightening gold coins by shaking them in a chamois bag which "sweated" minute particles off, and in the Late Empire, the milled, beaded or serrated edge, was abandoned.

Hellenistic jewelry and antiquities were coveted by collectors. Like the nouveaux riches of all ages, wealthy Romans were sometimes duped by clever forgers. There were elaborate workshops in Rome where "Greeklings," many of them from Magna Graecia, turned out amazing copies of goldwork, engraved gems, statues and paintings. Rome was a vast museum of the world's treasures both genuine and counterfeit.

The Romans adored extravagant displays. The kinds of objects they admired ran the gamut from immense bronze statuary to diaphanous silks literally worth their weight in gold. Pliny the Elder listed some of the articles displayed by Pompey in his third triumph celebrating his victory over the pirates of Cilicia in 67 B.C. They included "a gaming board, complete with a set of pieces, the board being made of gold and silver and measuring three feet broad and four feet long...three gold dining couches; enough gold vessels inlaid with gems to fill nine display stands; three gold figures of Minerva, Mars and Apollo respectively; 33 pearl crowns with gold; a square mountain of gold with deer, lions and every variety of fruit on it and a golden vine entwined around it; and a grotto of pearls on the top of which there was a sundial." Pliny commented that the display of these treasures taken from Mithridates, king of Pontus, started a rage for pearls. A dangerous amount of gold went East to pay for pearls and the emeralds, sapphires and rubies the Romans were so mad about. These precious stones were so popular that those who couldn't afford the real thing settled for superb imitations—so good, in fact, that ancient bogus emeralds made by Roman craftsmen were sold as genuine well into the last century.

Roman goldwork mirrors Etruscan, Greek, Hellenistic, Eastern, Celtic and Germanic influences. Early goldwork is indistinguishable from that of the Etruscans on whom the new city-state was culturally dependant. Growing traffic with the luxurious cities of southern Italy introduced the Romans to the exquisite jewelry and golden fabrics of Magna Graecia. Roman goldsmiths imitated Greek models without attaining the same level of craftsmanship. In general, Roman jewelry was produced for a clientele less concerned with refinement than show, whereas Greek jewelers worked for a more discriminating clientele.

After the Punic and Eastern Wars, Republican Rome came in firsthand contact with the Hellenistic kingdoms. The spoils of those wars, more than anything else, gave an impetus to Roman goldsmithing and jewelry production. From the East came elaborate jewelry in hundreds of forms emphasized richness of texture and variety of color. From the two commercial centers of Antioch and Alexandria came precious stones brought from India. From the East came gold brocades and huge, heavy gold-threaded tapestries as well as gossamer silks shot with gold threads. Nero once paid four million sesterces without grumbling for a gold-embroidered tapestry from Babylonia and 800,000 for a gold brocade from Alexandria.

Hellenistic jewelry achieved a synthesis of many styles—Greek, Persian, Scythian, Indian and Middle Eastern. Roman goldsmiths copied imports and observed foreign goldsmiths at work in Rome, some of them slaves. In the Late Empire new influences transformed Roman goldwork. As the Empire was subjected to disruptions on its frontiers, strong Oriental and Barbarian elements appeared in jewelry. The Oriental styles, originating in Egypt and Palmyra, featured large, heavy, overwrought pieces. The Barbarian styles reflect increasing Celtic and Germanic influences in the 4th century A.D. as the barbarian tribes besieged the northern boundaries of the declining Empire. Typically barbarian jewelry was large, graceless and strewn with glass paste, stones, such as garnets, and some gems. The gold served merely as a mounting, unlike Hellenistic jewelry in which gold and stones were equally important parts of an organic design.

Enough Roman jewelry has survived to confirm ancient literary descriptions. Yet, almost none of the large pieces, of which Romans had so many of, escaped destruction. Gold furniture, tableware and accessories on a large scale couldn't easily be buried in troubled times and fell prey to looters and the melting pot. We have to depend on poets and writers for descriptions of these vulnerable works that the Romans prized as an ideal way to advertise their worth. We can only imagine the extravagant displays of gold in the dazzling imperial city. Gold must have gleamed from roofs, facades, portals and massive statuary. Surfaces of wood, marble, masonry, even glass, were gilded. Domitian, in the 1st century A.D., spent enormous amounts of money to gild the great doors and roof of the temple on the Capitoline dedicated to Jupiter, Juno and Minerva. In the 5th century a Roman general stole the doors to pay his troops and the Vandals stripped the roof of its gold. The temple once boasted a spectacular gold and ivory colossus of Jupiter around which a three-storied colonnade was constructed. Gold was used so profusely to embellish the buildings of Rome that even in the Middle Ages when the city was but a shadow of its former self bits of gold were still visible.

The Greeks had invented the fire-gilding of masonry and sculpture in the mid-1st millennium. The Romans perfected the technique in which powdered gold was combined with mercury and spread on the surface to be coated. The surface was then heated until the quicksilver vaporized leaving a thin, uniform layer of gold. The Romans were aware of the connection between working with mercury and such afflictions as loss of teeth and hair, nerve degeneration and eventual death, but continued to practice fire-gilding, and Rome shimmered with gold inside and out. Today, amalgamation—as it is also called—is outlawed in most countries because of mercury poisoning.

Surfaces were also covered with beaten gold, both thick sheet and whisper-thin gold leaf. Roman goldbeaters excelled at production of gold leaf, beating an ounce into more than 750 sheets measuring "four fingers each way," according to Pliny the Elder. It was used for traditional applications as well as in making a novel kind of glassware. The Romans made very fine glass and developed a method of sandwiching gold medallions between layers of glass such as those roundels found embedded in the mortar sealing 4th-century tombs of early Christians in the Catacombs near Rome. A layer of gold was etched with religious scenes and then fused between sheets of glass forming a vessel or bowl. At the owner's death the medallion was removed along with its glass covering and used to decorate the tomb.

The goldsmiths of Rome clustered their shops in several areas such as the Via Sacra and belonged to various guilds dedicated to their specialized crafts. The *caetores*, for example, were metal chasers or engravers; *brattearii* made sheet gold and gold leaf; *auratores* or *inauratores* were gilders. A fresco found in the House of the Vettii at Pompeii is an interesting illustration of a sophisticated Roman goldsmith's atelier. It depicts plump, winged cherubs as goldsmiths and their apprentices who are casting, riveting, making gold chains, burnishing gold surfaces and enhancing goldwork with filigree, repoussé and engraving.

Cicero in his speeches against Verres, the *praetor* who made his fortune in Sicily, provides valuable information on Roman goldsmithing. He tells how Verres set up a large goldsmith's workshop at Syracuse where he had the art works he had plundered transformed by goldsmiths. Cicero says a vast number of men worked for eight months making nothing but gold dishes, vases and bowls on which to affix the *emblemata* or relief medallions, that had been removed from gold censer and sacrificial pieces the notorious governor had stolen. "The same man who now tells you he was responsible for the preservation of peace in Sicily," stated Cicero, "used to spend most of the day sitting in his workshop, dressed in a dark tunic and mantle."

Rome's contributions to the goldsmithing repertoire were limited to the elaboration of two previously invented techniques. They perfected the pierced goldwork known as *opus interrasile*. In the 3rd century A.D. this method of treating a gold surface like lace openwork became popular. It was probably of Syrian origin and remained fashionable during the Byzantine era. When it wasn't overloaded with stones or coarsely made *opus interrasile*, jewelry was quite beautiful. Roman goldsmiths also made advances in the niello work originated by the Mycenaeans. This was a type of champlevé enameling in which a fusible alloy made of various combinations of metals was used to fill cavities etched in a gold surface with a pattern in black rather than polychrome.

Neronian Rome reached a pinnacle of golden grandeur and vulgarity. As a modest teenage emperor he convinced the Senate not to erect life-size statues of him in gold and silver. Within a few years, however, he embarked on a career of the greatest cruelty and degeneracy. Even the relaxed moral climate of 1st century A.D. Rome was chilled by his deeds. Nero had enemies and innocents slaughtered at a whim, burned Christians as human torches, killed his tutor and wife and mother. His second wife was expecting a child when he killed her with a kick to the belly. He seemed genuinely grieved by her death and took as his third wife a young boy who he said resembled her, marrying him in a great public ceremony.

Nero's wedding gift to his peculiar bride was the *Domus Aurea*, the famed Golden House which inspired countless palatial mansions staffed by hundreds of slaves. The walls of the huge palace were plated with gold and stones; ceilings were coffered in ivory and gold; furniture was plated with gold, silver, ivory, gems and tortoise shell. Gold purchased statuary from Greece, paintings, ornate gilded candelabra, gold covered vases of Corinth bronze, and all manner of golden cloths and tapestries. The imperial galleys sheathed in gold floated on a pool "more like the sea." The palace complex was set in a mile-square fantasyland of meadows, gardens, and lakes. Nero built the Golden House and many other public buildings following a firestorm that raged unchecked through the city in 64 A.D. Nero, who was out of town at the time, was accused of having started the blaze. Legend still has it that he watched from a tower and fiddled as Rome burned. There is no doubt he was delighted to have the opportunity to undertake his construction project in which he spared no expense. He presided over banquets in his new palace where the flowers alone sometimes cost as much as the monthly food budget for the main palace. A giant gilded bronze statue of the emperor 120 feet high gave its name to the Colosseum which was later built near the site of the *Domus Aurea*.

The treasury was emptied by Nero's schemes and excesses. He confiscated the wealth of fellow Romans and wrung the provinces without mercy for more taxes. He increased the tribute due him but never had enough gold to meet his expenditures. Although gold was coming in from many sources, Nero sought new supplies. The year after the great fire he received a man named Bassanus who came from Carthage to tell the emperor of a dream in which he had been told by Dido where immense golden treasures lay hidden from the founding of Carthage. He said the gold was deep in a cave on land he owned and was in the form of "the shapeless and ponderous masses of ancient days" rather than in coins. According to Tacitus, who told the story about the emperor he loathed, Nero without questioning the report sent out the fastest triremes in his fleet to bring back the hoard. "Nothing," wrote Tacitus, "at that time was the subject of so much credulous gossip."

Nero also sought gold from less fanciful sources. Under Augustus expeditions had been dispatched to vanquish the gold-rich countries of Ethiopia and Arabia. They met with scant success and Nero decided to renew efforts to procure African gold. He dispatched an expedition that penetrated closer to the headwaters of the Nile than any Europeans for another 1,800 years. But they came back empty-handed.

Another area which attracted his attention was the auriferous zone of Central Asia beyond the Black Sea. Following defeat of the Achaemenid dynasty in the 3rd century B.C., Persia had been ruled by Alexander's chief heir, Seleucis, and his descendants until the Parthians, a people with origins in Central Asia, established feudal control over the area for five centuries. The Parthians were a thorn in Rome's side because they effectively monopolized the silk trade with China through control of the silk route. They grew enormously rich and kept Rome from direct access to the gold of the Caucasus and Bactria, the ancient country between the Hindu Kush and the upper Amu Darya (Oxus). As Nero's plans for an alliance with the Parthians never materialized, Rome continued to receive only irregular supplies of the Asian gold.

Nero was greatly cheered when a new vein of gold was found in Dalmatia which furnished him with another 50 pounds of gold a day to squander. When no amount of gold seemed to suffice, he turned to debasing the coinage. Nero permanently reduced the weight of the gold aureus although he didn't tamper with its purity. Following his example, other emperors tampered with their coinage in a vain attempt to stretch the supply. The silver denarius, which was the standard coin, had been worth about an ounce of silver in Augustus' time. It fell about 25% under Nero, and 200 years after the founding of the Empire was worth less than a fiftieth of its original value. The denarius was sometimes as much as half base metal. By the mid 3rd-century A.D. the Empire was flooded with almost worthless copper coins thinly washed with silver, over which gold was unable to exercise a steadying influence. If short-sighted rulers had resisted the temptation to debase Roman currency and if they had admitted the real value of the denarius, establishing a fixed ratio of silver to gold, the Empire's economy might not have become such a shambles with gold pouring out to pay for luxuries imported from areas where no other form of payment was acceptable.

Rome was plagued by a disastrous balance of payments for much of its history. Little of the gold that flowed into the heart of the ancient world stayed there. Although a tremendous amount of gold was immobilized in the form of plate and ornament, a great deal more was coined and went to pay for such exotic goods as blond slaves; pearls and oysters from Britain; furs and amber from the Baltic; perfumes, incense and cosmetics from Arabia; gemstones and jewelry from India; and ebony, ivory and wild beasts from Africa to name but a few of the things the Roman elite considered necessary for the good life.

Unfortunately, Rome never developed strong domestic industries. Eventually the provinces, which had been so cruelly stripped of wealth and exploited for the benefit of the capital city, began to develop prosperous manufacturing and agricultural industries. They exported their goods and began to absorb some of the gold and luxuries that had previously all gone to Rome. The provinces became vital, self-sufficient centers providing the Empire with emperors, generals and intellectuals in increasing numbers as the capital's star faded.

Incredible amounts of gold and silver were required to keep the vast Empire humming. The Elder Pliny reckoned annual currency payments to the East for Chinese silks and Indian gems amounted to more than 20,000 Roman pounds of gold. Although the drain eastward ebbed a bit after such imperial products as Syrian glass, Italian bronzes and the wine and oil of Asia Minor were offered as partial payment for imports, gold remained the chief form of payment. Roman exports couldn't begin to offset the volume of imported luxuries. Particularly after the 2nd century A.D. when Roman power and prestige began to decline, foreign countries were loathe to accept anything but gold. The silver currency was watered down so often, partly because silver deposits under Roman control were severely depleted, that not even Italians were willing to accept denarii.

The single greatest source of Roman wealth was Spain. The first Spanish gold came from the coastal zones of the east and south where gold had been washed from the rivers since prehistoric times. Rome gained easy access to the gold of Central Spain by the mid-2nd century B.C. In the area of the Guadalquivir River gold was so abundant that complex mining operations were unnecessary. Strabo, who visited placer mines in such Iberian rivers as the Douro and the Tagus in the 1st century B.C., described how men and women panned for alluvial gold. He also observed that gold dust and nuggets were found in arid spots by washing the sand with water channeled to the spot.

In 31 B.C. northern Spain came under Roman dominion and Augustus organized mining there as a state monopoly, setting the pattern which was followed by succeeding emperors. Roman mining engineers, the world's best until modern times, were trained at a state-run school in Spain. They learned prospecting methods and learned to detect auriferous ore by the presence of quartz gravels or copper ores. They were taught the techniques of sinking shafts, propping galleries with wood pillars, heating and splitting of the quartz matrix to release gold ore, drainage, assaying, refining and cupellation, the very ancient salt and sulfur process of refining gold. Apprentice engineers were introduced to a Roman invention of the 1st century B.C. called liquation which was a method of separating worthwhile amounts of gold from copper ore during the smelting process. This technique added considerably to Spanish gold production. Engineers also learned how to use mercury to separate gold from its crushed ores in the amalgamation process.

Spain continued to yield such a colossal amount of gold through the development of hydraulic mining. With this technique the mines of northwest Spain yielded amounts of gold comparable to the richest goldfields of the Egyptians, despite the low grade of the ore. Hydraulic mining is a quick, effective, but devastating method of using water channeled through complex works to break down relatively soft rock beds and expose gold bearing earth. Hydraulicking or hushing, as it was called in California where it was used in the 19th and early 20th centuries before being declared illegal, was an ideal way of processing previously impossible volumes of low grade ore. But it depended on a good supply of running water such as the Romans encountered not only in Spain, but in Gaul and Wales where they also introduced the practice.

In Spain, canals were dug leading from rivers to enormous

man-made reservoirs scooped out of a site overlooking a proposed mining area. The water was discharged under pressure, in great jets controlled so that it might fall as far as 800 feet onto the earth below, crushing the rock and freeing gold which was washed into sluices in drainage channels. The sluices were lined with shrubs such as rosemary to trap the gold particles as gravity pulled the detritus-laden water into streams leading to the sea. The larger bits of gold were removed by hand from the foliage. At intervals, the shrubs were removed, dried and burned. The ashes were washed to separate any gold dust.

The environment was irreversibly damaged. Whole mountains were washed away and rivers turned into drains carrying tons of debris to the sea which caused harbors to silt up and turned seaports into inland towns. But the price seemed small to the Romans who extracted some 20,000 Roman pounds of gold a year from the northwest goldfields, according to Pliny. He described in detail how gold in Spain was won in three ways—by washing, shaft mining and by "the destruction of mountains."

The third method of mining may seem to surpass all other achievements. By the light of lanterns mountains were hollowed out by galleries driven deeply into them. Lamps were also used to measure the length of the miners' shifts, for many of them would not see daylight in months. Because fissures could suddenly open up and crush the miners, arches were left at frequent intervals to hold the mountains up. In these mines—and in the shaft mines too—the miners broke up flint by using first fire and then vinegar. More often, because the galleries thus become full of suffocating steam and smoke, they broke it with iron rams of 150-pound weight. The pieces were carried out, night and day, in the darkness along a human chain. Only the last in the chain saw daylight. When all was ready they cut the "keystones" of the arches, beginning with the innermost. The earth on top subsided, giving a signal to a solitary look-out on a peak of the hill. With shouts and gestures he ordered the mine to be evacuated, and he himself sped down as well. The mountain then broke and fell apart with a roar that the mind can hardly conceive and with an equally incredible blast of air.

Diodorus aptly called the mines "golden prisons." Rome's gold was won at a high price reckoned in terms of human suffering and life. But since the miners were prisoners of war and slaves who worked naked and malnourished in the stifling dark until they perished, the cost was considered insignificant. Throughout the empire human skulls and bones have been excavated from ancient Roman mines along with the iron rings driven into subterranean gallery walls where shackled miners slept.

Since earliest times, miners had been faced with the problem of underground water seeping into the diggings. The Greeks designed drainage methods, but since Classical Greece regarded the practical applications of technology as unworthy and vulgar, they didn't test them properly. The Romans were first-class pragmatic engineers who adapted Greek designs and made great advances in solving problems. Roman slaves bailed with baskets woven of grass and covered with tar until engineers introduced the screw pump which the Greek Archimedes had developed for use in the mines of Sicily. Finally, Roman engineers improved the effectiveness of the cochlea and water wheel, more costly devices used in especially productive mines. Until the late 18th century A.D. Roman drainage solutions were unsurpassed.

Dacia, after Spain and Egypt, supplied the greatest supply of Roman gold. By the end of the 1st century A.D. the rich placers of central and southern Spain were showing signs of exhaustion. Production had fallen sharply from the previous annual yield of 10,000 pounds a year. The goldfields of Gaul also began to furnish less. Gold was more essential than ever to pay the armies, secure the vast frontiers and pay for the northern and oriental luxuries which had become essential to Roman life. The emperor Trajan, born in Spain and the first ruler from the provinces, gave Roman gold supplies a tremendous boost with his conquest in 106 A.D. of Dacia, comparable to modern Romania and Hungary.

Trajan was a great emperor, one of the few who was too upright to bribe the Praetorian guard. He had great ambition coupled with remarkable ability. Dreaming of a mighty Eastern empire like Alexander's, Trajan marched on the Parthians, who had twice defeated Roman armies, and took dominion over Mesopotamia, Assyria and Armenia, which still produced some gold, although the Pactolus river no longer brought down fabulous riches as it had in Midas' time. Trajan also secured the province of Arabia Petraea. Rome then controlled vital trade routes and collected import-export taxes, sales taxes and tithes from the trade in gold, precious stones and spices from Arabia and India.

But by far the greatest prize was Dacia which gave Rome a badly needed injection of precious metal. Its colossal royal treasury was the immediate trophy. Some accounts put it at five million pounds of gold and ten of silver. These ancient figures seem incredibly high and it has been suggested they represent ten times the actual spoil. Even so, half a million pounds of gold and a million pounds of silver would have been splendid prizes—enough to coin 22,500,000 gold aureii and more than 90 million denarii. Dacia's treasure was the last great war booty to come to Rome. Sources of new plunder were diminishing and by the mid 2nd century A.D. the long era of foreign conquests was all but over. Trajan returned from his brilliant campaign and was accorded a triumph lasting 123 days, scenes of which can still be seen carved on the arch bearing his name. He used the treasure to fund the greatest public works program since Augustus' reign, building harbors, roads and aqueducts. There was so much treasure that he handed out more than 100 denarii to every Roman citizen who applied for the donative and he sponsored four months of reveling during which 10,000 gladiators are said to have fought to amuse the mob.

The gold mines of Dacia proved far more valuable in the long run than the booty. The Carpathians had produced gold from prehistoric times and their potential was well known to the Classical world. Under strict Roman control the mines were run by a state appointed procurator who let out mining contracts on a percentage basis. The gold from the Transylvanian mines of Dacia added so much gold to the imperial supply that the price fell three per cent between 97 A.D. and 127 A.D., in spite of strict government regulation. To make up for the influx of gold and to keep the ratio of silver to gold fairly steady, fundamental to maintaining Rome's bi-metallic monetary system, Trajan increased the amount of base metal in silver coins.

No new sources of mineral wealth comparable to Spain and Dacia were found and the old sources began to play out. At the same time prospects for plunder faded. In fact, as the Empire was forced to withdraw from many areas it lost the gold mines of Transylvania. Rome also lost Gaul, Spain and Britain along with their supplies of raw gold until Aurelian restored them to imperial rule toward the end of the 3rd century A.D. Rome, the proud and ancient capital, ceased to be the heart of the Empire as the provinces gained vitality and wealth, absorbing quantities of gold, and furnishing most of the members of the Senate.

The military focus shifted to the northern and eastern borders where barbarian forces hammered away at thin Roman defenses. Subsidy payments to the barbarians accounted for a fatal drain on Roman gold stocks in the Late Empire. Domitian, in the 1st century A.D., set a precedent by giving gold to the troublesome Dacians before Trajan's conquest. Other payments followed, and as Roman expansion slowed and the various tribes learned to join forces to attack the most vulnerable points, more and more gold was being handed over. At the beginning of the 3rd century, bribes to the barbarian extortionists equaled the annual payroll of the Roman armies. Gold coins and specially minted gold bars from the Balkan mint at Sirmium accounted for much of the payment. The lion's share was in magnificent gold medallions bearing the emperor's likeness. These, at least, allowed the emperor to save face, since he presented them to the hovering tribal chieftains as "gifts." But no one was fooled. If the Roman payments were late the barbarians swooped across the borders to collect. Roman citizens began to abandon areas of Central Europe under continuous assault.

Many barbarians settled on Roman land and enlisted in the Roman armies. By the end of the 4th century the men and officers of the western army were almost all Barbarians who had either enlisted or been bought from tribal chieftains for gold.

Another drain on Roman gold was payment to the soldiers who proclaimed the "barracks emperor" during a half century of revolutions, when one infamous character after another bought himself the throne. Septimius Severus, one of these emperors, gave whispered advice to his sons Caracalla and Caligula as he lay dying: "Make your soldiers rich and do not bother about anything else." During the 80 years following his death in 211, every emperor died violently, as one ruffian after another murdered for the throne. Because an emperor knew his survival depended on the army's favor, there were frequent pay raises and gifts of splendid gold medals. Often unable to keep their lavish promises, these barracks emperors continually debased the coinage until the Roman denarius was unacceptable in marketplaces everywhere. The government was forced to accept taxes in grain and produce, and people reverted to wide-scale barter. Even the golden aureus, pride of the monetary system, was reduced by a third of its weight and sometimes adulterated until every coin was tested on scales and touchstone.

Rome was no longer the capital in any real sense. All of Asia was lost in the 3rd century while Italy was ravaged by pestilence and famine and the barbarians stepped up their attacks on every border. Modern finds of Roman gold hoards buried in the Balkans, Greece, Italy and Asia Minor are evidence of barbarian attack. Despite several attempts at monetary reform, the situation was out of control. Prices soared and people hoarded what gold they had in the face of insecurity and crushing taxes. Inflation devastated the economy, paralyzing trade and industry.

The Roman Empire was dying. There were a few brilliant flashes during the centuries before the fall which echoed her former grandeur, but the end was inevitable. Aurelian, hailed as "Restorer of the World," provided a respite by driving back the enemy everywhere but along the Danube. As Germanic tribesmen plundered Italian cities along the Po River and menaced Rome, so the emperor set all the city's guild members to work constructing the amazing brick wall encircling Rome, parts of which still stand. He claimed to be divine, the earthly representative of the sun, and used the sun's metal to dazzle Romans and Barbarians alike. His court was that of an eastern god-king. Men knelt worshipfully before him as he sat on a throne of beaten gold, resplendent in furs, gems and ornaments of exquisitely worked gold. For all its pomp and glitter, Aurelian's gorgeous court failed to disguise the rot that was undermining the Empire.

Two able emperors in the late 3rd and early 4th centuries, Diocletian and Constantine, did what they could to strengthen the empire. In 284 Diocletian, son of a former slave, officially abandoned Rome and set up court in Asia Minor where he could more easily deal with threats from Sassanian Persia. His new capital near Byzantium was rebuilt under Constantine, the first Christian emperor. As Constantinople, capital of the Byzantine Empire following the fall of Rome, the city became the focus of the world's gold.

Diocletian ruled in the tradition of an oriental despot. From his golden throne he issued edicts to courtiers who kissed the hem of his robe. His absolute rule enabled him to carry out reforms aimed at curing the Empire's ills. An elaborate bureaucratic organization and a complex espionage network aided his efforts. He tried several methods of improving the currency. His most outstanding contribution was issuing gold coins of reliable purity and weight. He minted the gold solidus which weighed a sixtieth of a pound. Constantine reduced it to a seventy-second of a pound. The solidus and its third, the tremissis, were among the world's longest circulating coins. They were accepted in Europe, Asia and Africa until the fall of the Eastern Empire in 1453. The Vandals, Lombards, Ostrogoth and other barbarians who fought over the corpse of the Western Empire, used Diocletian's coins as models for gold coinage of the Early Middle Ages.

In the 7th century, when the Arabs wrested Syria from the Byzantines, they adapted the solidus as their unit of gold currency. Since the Koran forbade portrayal of the human figure, the bust of the emperor was replaced with Muslim designs. The dinar, still in use in the Arab world, got its name from the denarius aureus, the "gold denarius."

Diocletian's gold coins, good as they were, could only be minted in limited quantities because gold was scarcer than ever. The state was desperate for gold to pay the barbarians and maintain its armies. Every possible means was used to wring gold out of people. Cities were forced to sell their gold reserves to the state and certain taxes, formerly payable in kind, were demanded in gold. Constantine carried on Diocletian's reforms, adding to imperial revenues by continuing a tax levied on goldsmiths and other craftsmen as well as merchants, which was payable only in gold. Corrupt tax officials made huge fortunes by siphoning off gold that should have gone to the treasury. The military looted Roman cities, taking whatever they wanted from citizens under the guise of requisitions. Military officers refused payment in anything but gold and military costs skyrocketed. It was understood that without the armies the Empire had no chance of survival.

There still were some wealthy men. Those who had made illicit fortunes in the civil service used their gold to buy up vast estates at a time when many formerly prosperous middle class Romans were reduced to poverty. There were senators, according to ancient sources, who, even in the closing years of the Empire, had annual incomes in gold of up to 4,000 pounds. These were most often men who had cleverly spread land holdings throughout the Empire and didn't lose everything at once.

Because gold was in hiding, less gold was in evidence than at any time since the founding of Rome. Small landholders were ruined by taxation, debt and the predations of outlaws, whose numbers swelled as the situation became increasingly chaotic. Diocletian came to power in a world exhausted by violence and disorder. He made vigorous efforts to stabilize conditions. With the power of a despot he practiced strict control over wages, prices and the destinies of farmers and craftsmen who were tied to their lot in life and bound to follow their fathers. If a bankrupt farmer was forced to sell, he found himself a serf, tied by law to the land.

Under Constantine life became even more intolerable and Roman citizens actually left to live in barbarian areas. Evidence of increased communications between Romans and barbarians appears in the imperial prohibitions on trading gold across the frontiers. Barbarian tribesmen pillaged and plundered in formerly secured areas. The armies which fought them off were largely Barbarian in makeup and demanded higher wages, paid in gold, which paralleled rising barbarian demands for golden payoffs.

Constantine may have been moved to convert to Christianity because he then could confiscate treasure from the thousands of pagan temples in his Empire. After 331 A.D. the imperial mints suddenly began to pour out a greatly increased volume of gold coins but even this wasn't enough. Wealthy men, or men known to have gold, were compelled to present "gifts" of gold to the emperors. Valentinian, emperor in the late 4th century, for example, received 1,600 pounds of bullion and coin as birthday presents.

The Roman Empire was so far gone by then that no amount of gold could save it from the wild horsemen who came out of the north. The old spirit of liberalism and patriotism, the Greco-Roman ideals, were long gone. The brilliance and energy that had shaped the world's greatest Empire had given way. Barbarian forces took the Western Empire bit by bit. In 410 A.D. the Eternal City which had resisted invasion for almost a thousand years was ravaged by the Visigoths. Two payments of gold and peppercorns which were equally precious, couldn't save the city. The final blow came in 455 when marauding Vandal pirates sacked Rome. In 476 A.D. the last Roman emperor of the West, Romulus Augustulus, stepped down to let Odoacer, a German selected by fellow officers, assume the crown. Romulus, who ironically bore the names of Rome's legendary founder and her first emperor contented himself with a generous bribe of gold and quietly retired to the country.

Civilization had never been dealt such a cataclysmic blow. With the fall of Rome the West was plunged into a maelstrom of darkness as Germanic tribes, Magyars, Arabs and Norsemen swept back and forth across the face of Europe. In the so-called Dark Ages which followed the collapse of the Western Empire gold bypassed the spent city of Rome for Constantinople, the brilliant Byzantine capital whose golden luxury has passed into legend. The Eastern Empire had been able to promise Attila the Hun 2100 pounds of gold a year to keep "the Scourge of God" and his hordes from seizing the city.

At its peak the Byzantine Empire attracted the world's gold production and paid in gold for Chinese silks, sumptuous Persian carpets, Russian furs and the jewels and spices of the East. Meanwhile Greco-Roman civilization in Europe disappeared along with the end of city life. Trade and industry ground to a halt with the passing of Rome's able organization and administration. The excellent network of Roman roads linking every town fell into disrepair. Bridges and aqueducts crumbled; the art of bricklaying was lost. Flourishing urban centers no longer supported a gold-loving population of aristocrats, prosperous merchants and talented craftsmen. Cosmopolitan Greco-Roman civilization, which valued gold as an index of economic and political power, was replaced by an agrarian society with clusters of frightened people huddling in self-sufficient settlements.

Roman technical knowledge was lost. The mines were abandoned. There was no mining organization, there were no good roads for transport and, above all, there were no markets to absorb the gold, work it and distribute it. The rivers of Spain, France and the Alps supported free lance prospectors just as they had thousands of years earlier. Men washed surface ores in a haphazard fashion and some gleaned the mounds of tailings from Roman workings. Heedless of the danger from crumbling galleries other men braved the abandoned mines filled with water. They found some gold but the barbarians couldn't begin to match Roman skill at deep and hydraulic mining.

At the time Rome fell, world stocks of the eternal metal were at a record high. Men had been accumulating gold for more than 4,000 years. Yet from the 5th to the 12th century, Rome, which had blazed so bright with golden glory, was starved. There was some gold of course in Medieval Europe, but it was once again the prize of kings and God. The beautiful metal that so many ordinary men and women in the Roman world had used to display their position or enhance their attractiveness was unavailable. European royalty and the powerful Christian church, protected and nurtured by the kings and queens of Europe, accounted for almost all gold ownership in the 700 years following the collapse of the Western Empire.

Gold is an amazing metal. The word "gold" itself conjures up a kaleidoscope of fascinating images from the history and lore of every age. Since the dawn of time it has provoked both the noblest and the basest of behaviors. No other substance has served as such a catalyst in the rise and fall of civilizations as the inert metal which always emerges still gleaming and in no way diminished from whatever drama it has precipitated. In the ancient world, gold began and ended as the divine metal. In the modern world gold, no longer sacred, decorates the drinking glasses of hundreds of millions of people. The lure of gold is as old as mankind and, in later times, the search for both real and legendary golden treasure was to draw men through unknown lands and across uncharted seas to discover new worlds. Africa, China, Japan and the entire Western Hemisphere were explored by adventurers in the centuries after the European darkness had faded.